Intelligent Automation with End-User Computing Solutions

A Guide to Empowering Efficiency

Ajit Pratap Kundan

Apress®

Intelligent Automation with End-User Computing Solutions: A Guide to Empowering Efficiency

Ajit Pratap Kundan
Faridabad, Haryana, India

ISBN-13 (pbk): 979-8-8688-1311-5   ISBN-13 (electronic): 979-8-8688-1312-2
https://doi.org/10.1007/979-8-8688-1312-2

Managing Director, Apress Media LLC: Welmoed Spahr
Acquisitions Editor: Aditee Mirashi
Coordinating Editor: Jacob Shmulewitz
Copy Editor: Kim Wimpsett

Cover designed by eStudioCalamar

Cover image by Unsplash.com

Distributed to the book trade worldwide by Springer Science+Business Media New York, 1 New York Plaza, New York, NY 10004. Phone 1-800-SPRINGER, fax (201) 348-4505, e-mail orders-ny@springer-sbm.com, or visit www.springeronline.com. Apress Media, LLC is a Delaware LLC and the sole member (owner) is Springer Science + Business Media Finance Inc (SSBM Finance Inc). SSBM Finance Inc is a **Delaware** corporation.

For information on translations, please e-mail booktranslations@springernature.com; for reprint, paperback, or audio rights, please e-mail bookpermissions@springernature.com.

Apress titles may be purchased in bulk for academic, corporate, or promotional use. eBook versions and licenses are also available for most titles. For more information, reference our Print and eBook Bulk Sales web page at http://www.apress.com/bulk-sales.

Any source code or other supplementary material referenced by the author in this book is available to readers on GitHub (https://github.com/Apress). For more detailed information, please visit https://www.apress.com/gp/services/source-code.

If disposing of this product, please recycle the paper

Table of Contents

About the Author

Ajit Pratap Kundan stands at the leading edge of the most innovative cloud technology in today's world. He has helped position VMware as a leader in the private cloud area through an SDDC approach. An innovative presales techy with 22+ years of tech industry experience, he has promoted cloud technologies to his government and defense customers. Ajit is a valued writer on cloud technologies.

About the Technical Reviewer

 **Ameya** is a senior consultant with a decade of experience who specializes in developing ETL solutions for databases using Oracle PL/SQL and VMware products. He has worked on projects that enhance user experience and data security and has a keen interest in AI, data quality, and data governance. In his free time, he enjoys reading, playing outdoor games, and spending time with his family.

Introduction

This book is targeted at a specific audience interested in automation, VMware technologies, and IT infrastructure. If you want to learn to use VMware automation tools, improve operational efficiency, or learn advanced features for virtualized environments, this book will likely be a valuable resource. Here's a breakdown of who might benefit most from reading this book:

- IT professionals and system administrators
  - Individuals responsible for managing VMware environments
  - Professionals seeking to streamline their daily tasks with automation
  - Those looking to enhance their knowledge of VMware vSphere, vRealize, or similar tools
- Cloud and infrastructure architects
  - Architects designing and implementing virtualized and cloud environments
  - Professionals aiming to integrate VMware with automation and orchestration platforms

- DevOps engineers
  - Those focusing on infrastructure as code (IaC) practices
  - Engineers wanting to leverage automation tools to manage and deploy VMware environments efficiently
- IT managers and decision-makers
  - Managers exploring automation solutions to reduce operational costs and improve efficiency
  - Those evaluating VMware's automation capabilities for their organization
- Students and enthusiasts
  - Individuals eager to learn about VMware's automation ecosystem
  - IT students looking for practical insights into intelligent automation concepts
- Consultants and IT service providers
  - Professionals offering VMware-based solutions who need to stay updated on automation trends
  - Consultants looking to design and implement tailored automation strategies for clients

CHAPTER 1

Business Value of End-User Computing

In this chapter, we will discuss a desktop-virtualization solution from VMware as well as newer technologies built specifically for a mobile and collaborative workforce. We will cover how these technologies together enable IT to optimize its current environment while safely embracing innovation and emerging trends to maintain a productive workforce and a secure business environment. We also see how the VMware solution accelerates application deployment and simplifies application migration with agentless application virtualization and how it provides secure access to applications and data on any mobile device or computer, enhancing the end-user experience while reducing management costs. We will cover the products required to simplify the management, security, and control of desktops while delivering the highest-fidelity experience of desktop services to any device, on any network.

Purpose and Intended Audience

This chapter provides detailed information about the end-user requirements for the software, tools, and external services required to successfully implement the VMware end-user computing platform. Before you start deploying the components of the VMware solution, you

© Ajit Pratap Kundan 2025
A. P. Kundan, *Intelligent Automation with End-User Computing Solutions*,
https://doi.org/10.1007/979-8-8688-1312-2_1

must set up an environment that has a specific compute, storage, and network configuration and that provides services to the components of the software-defined data center (SDDC).

This chapter is intended for EUC architects, infrastructure administrators, and VDI administrators who are familiar with and want to use VMware software to deploy and manage an SDDC.

While the rapid influx of smart devices, including tablets and phones, and mobility workforce trends are adding significant complexity to IT operations, the increase in employee productivity, collaboration, and satisfaction that mobile technologies offer is not lost on business leaders.

Business leaders are increasingly looking to IT to deliver a mobile workspace that allows employees to access corporate data, applications, and communication resources on their devices of choice. But making the most of the opportunity while minimizing risks can be complex and overwhelming for many IT teams. Therefore, it makes more sense than ever for IT leaders to consider virtual client computing (VCC) solutions.

The benefits of VCC include centralized desktop and application management, "any device" access to corporate IT resources, an increased ability to protect corporate intellectual property, and demonstrated compliance with industry and governmental regulations.

However, when IT cannot optimize performance and effectively support the environment, this has a direct impact on the end-user experience, adoption, and ultimately success of the implementation. It's not just about saving money or improving efficiency. For VCC to be successful, the technology must provide an end-user experience that is as good as, if not better than, the end-user experience on local operating systems and applications.

Few organizations that have virtualized desktops and applications understand the impact on their costs, operations, and businesses. Research shows that these organizations are achieving substantial value with a VDI solution, and the solution will yield an average return on investment within five years by:

- Supporting business operations through employee mobility and access to applications

- Saving time associated with device and application log-ins

- Requiring less IT staff time to support and manage device environments

- Reducing the impact of device-related problems on users and the business

- Costing less than traditional PCs and other nonvirtualized devices

We will discuss the new post-PC era of technology, what IT approach is needed, and how customers are using validated solutions with VMware.

The Way We Collaborate and Work Is Evolving Rapidly

People want the type of experience they have in their personal life at work. They expect to be able to use any device they want and get their apps and data anywhere and also easily share information with their friends. This is an era of diversity called the post-PC era. The pace of technology change is faster than ever.

The post-PC era allows us to:

Simplify: Virtualize desktops and applications into the data center.

Manage and secure: Secure desktops in the data center and remove data off endpoints.

Empower: Give roaming access to desktops across devices and a familiar desktop experience.

The Need for Automation in Modern Business

End-user computing with automation empowers users to work smarter, not harder, by leveraging technology to streamline their workflows and optimize productivity.

Each customer journey is different. There is no single path, so prioritize the solutions important to each and every customer requirement.

People don't buy products; they buy solutions to their problems.

We must get all details on the products that align to a solution and help the customers meet their needs.

VMware Anywhere Workspace is aligned with a customer objective of "Any app from anywhere on any device " across verticals.

Build Trust for the New Distributed Workforce

Organizations must have an employees-first policy with device choice, flexibility, and seamless, consistent, high-quality experiences. They have to ease the move to zero trust with situational/contextual intelligence and connected control points. Customers need solutions that can manage outcomes, not tasks, with intelligent compliance, workflow, and performance management

- **Make intelligence work for organizations:** Organizations should take advantage of unified intelligence to quickly identify, prioritize, and generate automated actions.

- **Custom workflows made easy:** Customers can create and deploy complex workflows with simple drag-and-drop flexibility and speed.

- **Cloud-native visibility and management:** Proposed solutions can extend visibility and management to 100% of devices with cloud-native, off-domain policy management.

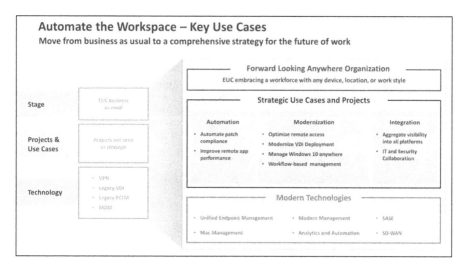

Figure 1-1. *Key use cases*

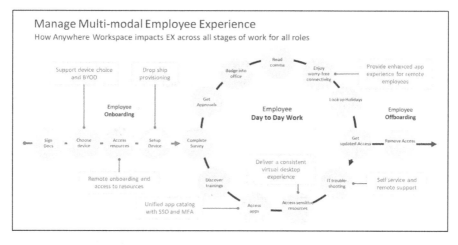

Figure 1-2. *Daily operations*

- **Touch-free IT:** Deliver frictionless experiences, from pre-onboarding of employees to proactive and instant support anywhere, anytime.

- **Freedom of choice:** Support knowledge and frontline employees working in any location using any device or OS for both bring-your-own (BYO) and corporate-liable devices.

- **Uncompromised uptime:** Ensure a good application experience whether working from the office or remotely.

- **Secure the distributed edge:** Anywhere Workspace secures the distributed edge. This enables the move toward end-to-end zero-trust security across the solutions.

- **Embrace least privilege:** Shrink the attack surface. For example, Workspace ONE Tunnel and Workspace ONE UEM can minimize the footprint of enterprise data on a device and in transit. This tunnel further integrates with SD-WAN and NSX for security all the way to the workload. By shrinking the attack surface with micro-segmentation and real-time continuous authentication and authorization for access control policies, we can secure all the endpoint devices along with data and apps.

- **Situational intelligence:** You can enable situational awareness, with visibility into devices, users, networks, and applications, as well as corresponding control points. Integrations between Workspace ONE and Carbon Black enable real-time response to threats. It

can further strengthen the authoritative context of your
environment and threat intelligence that is trustworthy,
actionable, and readily available.

- **Connected control points:** Visibility and control can
  be extended via Workspace ONE Trust Network, open
  APIs, and integrations with ITSM tools. Workload,
  network, device, and access controls, and teams that
  connect and align to the applications and data are
  being protected with these tools.

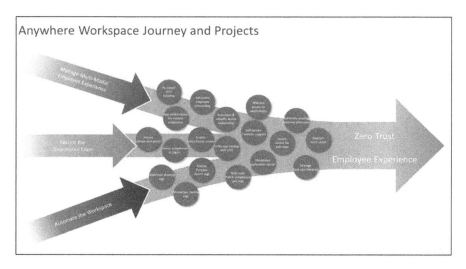

Figure 1-3. *Zero-trust employee experience*

Benefits to Customers

Internal alignment is crucial to make our customers successful. We need
common building blocks to help deliver consistent messaging, tooling,
PNP, PSO, competitive differentiations, and more. We should go with
smart discovery based on projects/business objectives, and priorities can
be tailored to customers and provide before/after assessments to realize
value. A model should be extensible for new projects as and if required.

Anywhere Workspace Business Objectives: Foundational Blocks for Solutions

Anywhere Workspace has the following building blocks.

Manage the Multimodal Employee Experience

This is the first building block of Anywhere Workspace.

1. Support a personal/BYO initiative.

2. Ensure app performance for remote employees (SD-WAN, Horizon).

3. Automate employee onboarding.

4. Automate existing business processes (e.g., SFDC, SNOW).

5. Unify the app catalog with SSO.

6. Automate and simplify device onboarding (e.g., drop ship).

7. Create VPN-less access to applications from remote locations.

8. Enable self-service/remote support.

Organizations should enable the Browser Only workspace on devices and extend their BYO policy to Macs devices. The proposed solution must enable remote collaboration to improve business continuity and resilience.

Secure the Distributed Edge

These are building blocks of Workspace anywhere.

1. Secure remote endpoints (Carbon Black).

2. Enforce automatic virus-scan/device compliance at logon.

3. Enable policy-based access.

4. Modernize application access (e.g., SAML versus AD).

5. Enable secure access for web apps (CWS).

Customers can migrate McAfee to Bitlocker and enforce behavioral risk controls.

They can reduce infrastructure cost for remote work (VPN) by enforcing data sovereignty.

Automate the Workspace

Part of building blocks.

1. Modernize desktop management (Windows/Mac).

2. Modernize mobile management (iOS/Android).

3. Enable special-purpose device management.

4. Automate patch compliance and management.

5. Support a multicloud environment.

6. Manage SaaS app life cycles.

Customers can extend the life of legacy applications. They can allow remote use of the "fat-client" WinApp from anywhere and also migrate the on-prem app workload to SaaS.

Automate and Simplify Device Onboarding

Part of building blocks.

- **Setup of new devices for users**

  - Day 1 for new employees: magic link experience

  - Device refresh for existing users

- **Automate deployment and configuration**

  - Windows, macOS, iOS, Android, ChromeOS

  - Windows and macOS provisioning, out-of-the-box enrollment

Use Cases	Project Combinations			
How do I plan for my distributed workforce?	Automate patch compliance and management	Modernize desktop management (Windows/Mac)	Enforce automatic virus-scan / device compliance at logon	Secure my remote endpoints
How do I ensure performance for employees on all devices?	Support personal/BYO initiative	Enable self-service / remote support	Ensure app performance for remote employees (SD-WAN, Horizon)	VPN-less access to applications from remote locations
How do I secure access from any device from anywhere around the world?	Enforce automatic virus-scan / device compliance at logon	Enable policy-based access	Enable secure access for web apps (CWS)	Secure my remote endpoints
What can I do to improve existing processes?	Automate employee onboarding	Automate existing business processes (e.g. SFDC, SNOW)	Automate and simplify device onboarding (e.g. drop-ship)	Automate patch compliance and management
How do I ensure my IT environment is setup for future success and capable of incorporating future technology?	Enforce automatic virus-scan / device compliance at logon	Modernize mobile management (iOS/Android)	Unify app catalog with SSO	Modernize application access (e.g. SAML vs AD)

Figure 1-4. *Solution with use cases*

- **Access to**

  - Applications and catalog

  - Training

  - SSO to apps and data

- **Managed and unmanaged device options**

    - Mobile, iOS, Android

Benefits

Benefits of above building blocks.

- Reduce IT time on manual device onboarding

- Imaging and application install

- Faster time to employee productivity

- Better employee experience

Workspace ONE Drop Ship Provisioning

Workspace ONE Drop Ship Provisioning is a "ready to eat" kind of solution designed to simplify and accelerate the deployment of purpose-built devices. Organizations can ensure that devices are fully configured, secure, and ready for use upon arrival, minimizing the need for manual setup by IT personnel.

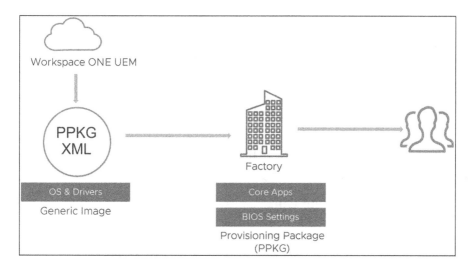

Figure 1-5. *Reducing manual efforts*

Offline mode improves the user experience because end users receive a fully operational device right out of the box, reducing downtime and increasing productivity. It also lowers opex costs and is ideal for large-scale deployments by ensuring consistent configurations across all devices. It also ensures that devices are compliant with corporate security policies from the moment they are powered on.

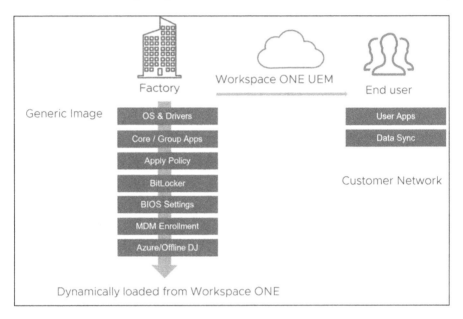

Figure 1-6. *Innovative solution for corporates*

Modernize Application Access

- Multiple ways of modernizing access to applications depending on application capabilities

- Integration with Workspace ONE Access

- SAML, SSO, Conditional Access

- VMware Tunnel

- Identity bridging with Unified Access Gateway

- Modern authentication into Horizon

- Then access traditional apps that cannot be natively SAML integrated

Benefits

- As apps migrate to the cloud, security policies need to evolve and with a distributed workforce, securing access to applications is vital

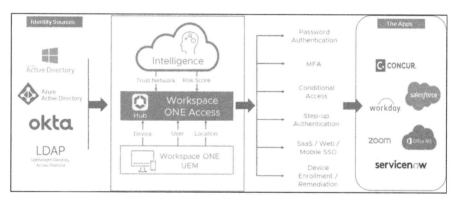

VMware Tunnel: Permit individual apps access to corporate resources and Secure traffic from device

Uses: Workspace ONE UEM ,Unified Access Gateway

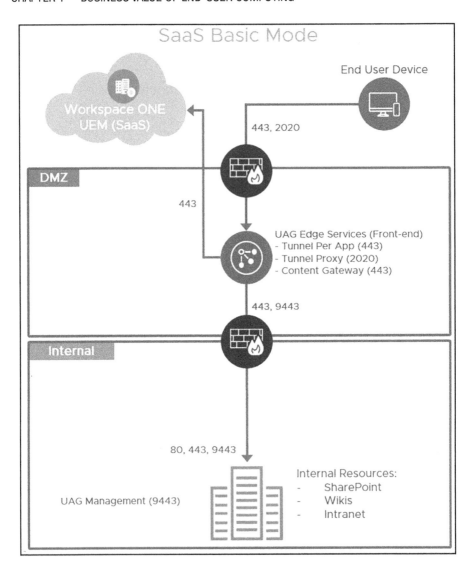

Figure 1-7. Identity Bridging with Unified Access Gateway

1. Client navigates to application URL

2. Client is redirected to the IdP (Workspace One) for authentication. The IdP issues a SAML assertion upon authentication.

3. Client passes the SAML assertion to Unified Access Gateway which then validates that the SAML assertion is from a trusted IdP.

4. Unified Access Gateway extracts the client's username from the SAML assertion and requests a Kerberos ticket from Active Directory on behalf of that user.

5. Unified Access Gateway authenticates against the internal web server using the Kerberos ticket obtained from AD.

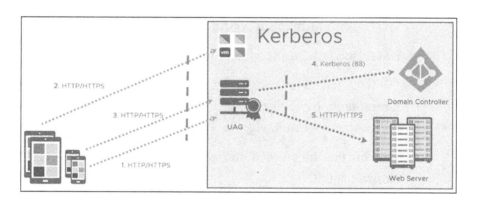

SAML/ Certificate to Kerberos

Modern Authentication into Horizon

This is to start the process with authentication.

1. Launch a Horizon resource from Workspace ONE Access.

2. Workspace ONE Access generates a SAML assertion and artifact.

3. A Horizon client is launched from a view URL.

4. Unified Access Gateway proxies the authentication to the Horizon Broker.

5. The broker performs SAML to resolve against Workspace ONE Access.

6. Workspace ONE Access validates the artifact and returns an assertion.

7. The broker returns a successful authentication.

8. UAG returns the successful authentication to the client.

9. A remote protocol client launches a session with the parameters returned.

10. UAG proxies the protocol session to the Horizon agent.

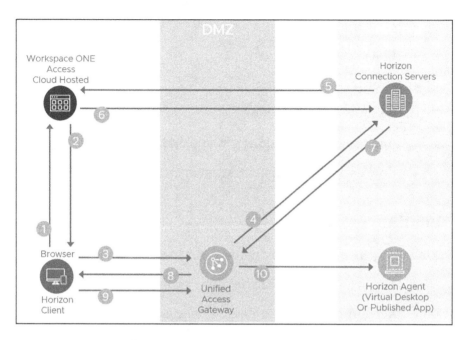

Figure 1-8. *Launching an application inside a Horizon session*

Modernize Desktop Management (Windows and Mac)

Desktop Management Process: To ensure secure, efficient, and standardized operation, maintenance, and lifecycle management of desktop (and laptop) systems in an organization.

- Cloud-managed, unified management of devices
- Windows, macOS
- Non-domain-managed
- Security
- Baselines (versus GPO)
- Custom settings

17

- Device compliance

- Bitlocker, sensors

- Over-the-air configuration

- OS updates and patches

- Application distribution (new and updates)

- Co-exist with traditional management (e.g., SCCM)

Benefits

The following are the benefits of modernizing desktop management:

- Managing the full life cycle of any endpoint device

- Keeping devices up-to-date on policies, patches, and newest versions

- Ensuring compliance and device health

- Setting up an over-the-air policy and configuration including off-domain

- Automating onboarding and reducing time and cost

Workspace ONE UEM

Workspace ONE Unified Endpoint Management (UEM) is a comprehensive platform designed to provide a seamless and secure management experience for all endpoints in an organization. It integrates various management functions into a single platform, allowing IT departments to manage desktops, mobile devices, rugged devices, and IoT endpoints from a unified interface.

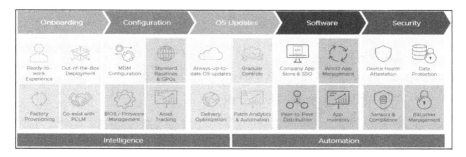

Figure 1-9. *One solution for all*

Workspace ONE streamlines the day-to-day operations by reducing the complexity of managing diverse endpoints with a single, integrated solution; ensures devices comply with corporate security policies; and mitigates risks associated with data breaches and noncompliance. It also helps to provide better user experience for end users with easy access to necessary applications and resources. It lowers opex costs by reducing the need for multiple management tools and increasing IT productivity and scales up/down easily to accommodate growing numbers of devices and users, supporting business growth and digital transformation initiatives.

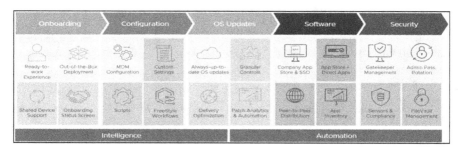

Figure 1-10. *Scale on demand*

Manage Multimodel Employee Experiences

Managing multimodel employee experiences involves integrating different modes of work (e.g., in-office, remote, hybrid) to create a cohesive and effective environment for all employees.

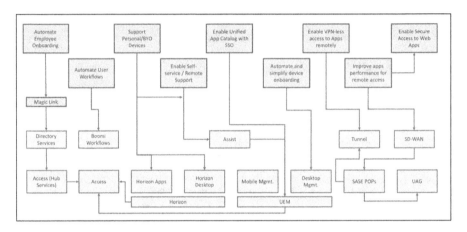

We have to acknowledge that one size doesn't fit all. We need to tailor experiences to different roles, teams, and individual preferences and leverage unified communication tools to implement platforms like Slack, VDI, or remote support software tools to ensure seamless communication across all work modes.

Secure the Distributed Edge

Securing the distributed edge can be achieved by implementing strategies and technologies to protect the data, devices, and networks located at the edge of an organization's IT infrastructure. This is critical as most of the devices connect to the network and also data is processed closer to where it is generated for precise outcome.

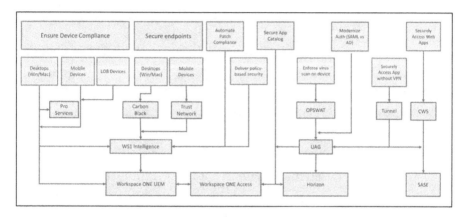

We have to isolate edge devices by segmenting edge devices from the core network to minimize the attack surface and also use micro-segmentation to create smaller, more manageable security zones within the network. We have to follow zero-trust security practices with identity verification by implementing strict identity and access management (IAM) policies. First verify all users and devices trying to access the network and then trust with least privilege access to ensure that users and devices have only necessary access to perform their functions.

Automate Your Workspace

Automating VDI can significantly improve efficiency, scalability, and user experience.

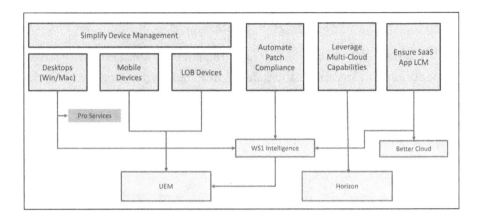

Self-service portals help users with self-service portals for day-to-day tasks like password resets or virtual desktop provisioning and automated support by implementing chatbots or automated scripts to handle common support queries and issues.

Using the web-based management platform, administrators can create customized sets of applications and data access (workspaces) for end users. This includes setting security policies and application entitlements. Using their desktops, mobile browsers, or mobile applications, employees can gain access to work resources, including shared corporate documents and applications. Enterprises can continue to manage and update the workspaces. If necessary, they can remotely remove an entire workspace, including all corporate data.

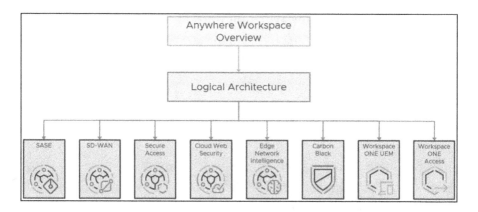

The following are the VMware Horizon workspace solution components:

Workspace Client: Install this component on endpoints to enable them to gain access to the workspace environment as well as to perform actions, such as file synchronization to the local device.

Workspace Configurator: This component is used at initial setup to provide a wizard for setting up the modules that will be used.

Workspace Manager: This gives you access to the Administrator web interface to be able to manage policies, user entitlements, and the global catalog. It is also referred to as the Service VA.

Workspace Connector: This component provides user authentication, directory synchronization, thinapp delivery, and VMware view integration.

Workspace Data: This component stores the user's files, as well as enables shared files and folders.

Workspace Gateway: This is the end-user entry point, used to route traffic to the correct appliance. Please use a friendly FQDN for end-user testing.

Summary

VMware Horizon Workspace is an end-user computer solution for organizations that want to simplify IT control while empowering the mobile workforce. Horizon Workspace simplifies the end-user experience and reduces IT costs by combining applications and data into a single enterprise-class aggregated workspace, securely delivered on any device. End users gain freedom of mobility through anytime, anywhere access. For the administrator, the results are simpler, centralized, policy-based management and control of IT consumption to improve the total cost of ownership.

CHAPTER 2

Empower Frontline Workers with Workspace ONE

In this chapter, we will explore VMware solutions for frontline workers and how these software tools can provide robust, secure, and efficient mechanisms and platforms to workers who are the first point of contact with customers or handle critical operations on-site. Frontline workers include retail associates, healthcare professionals, field service technicians, and manufacturing operators. VMware's solutions are designed to empower these workers with the technology they need to perform their tasks effectively and securely. We will also go through different use cases from various industry like retail, healthcare, supply chain, and manufacturing.

Purpose and Intended Audience

Before you start the project in this chapter, you must set up an environment considering real-time scenarios and how these frontline workers provides services to the components of the software-defined data center (SDDC).

© Ajit Pratap Kundan 2025
A. P. Kundan, *Intelligent Automation with End-User Computing Solutions*,
https://doi.org/10.1007/979-8-8688-1312-2_2

This chapter is intended for EUC architects, infrastructure administrators, and VDI administrators who are familiar with and want to use VMware software to deploy and manage an SDDC.

We can grow Workspace ONE market share by building awareness and driving demand for Workspace ONE UEM for frontline worker use cases with mission-critical devices used by frontline workers across retail, healthcare, and manufacturing industries.

Frontline Workers

A *frontline worker* is a service or task worker who isn't tied to a desk and who reports to a jobsite or is in the field. This worker relies on mission-critical devices to complete their tasks. Example occupations include delivery drivers, warehouse workers, store associates, and nurses.

Mission-Critical Devices

A *mission-critical device* is one considered essential to the success of a specific task or operation, so if it fails, it immediately impacts the company's bottom line. It is typically corporate-owned, shared by multiple workers and optimized to access line-of-business (LOB) app cases only. We as solution providers have to identify problems of frontline workers across business, replacing existing competitive solutions not able to solve business issues, and position Workspace ONE for mission-critical device management.

We as solution providers have to educate existing Workspace ONE customers to use UEM licenses for both knowledge and frontline worker use cases and make sure they're leveraging mission-critical device management features. We can create opportunity to upsell Workspace ONE Frontline Worker or Workspace ONE Assist add-ons by adding values to existing investment.

We as solution engineers or service providers have to identify knowledge worker and frontline worker use cases across business and position Workspace ONE UEM for mobile, laptop, and mission-critical device management.

IDENTIFYING OPPORTUNITIES

What kind of shift-based frontline workers do you employ on-site and in the field across your business? How are your workers using these devices across your supply chain? Are you using a single platform to manage knowledge and frontline worker use cases? Are you using a single platform across all frontline worker use cases? If siloed, what other solutions are you using and which teams handle what? What are the biggest challenges/ pain points you encounter with your existing solutions? Do these existing solutions meet your requirements for enrollment and configuration, scalability, device updates, shared device management, access, EX, analytics and automation, and remote support? Have you ever considered consolidating your management silos under a single UEM platform, like Workspace ONE?

Value Prop

Workspace ONE is built to address the unique management requirements of mission-critical devices at scale, with proven complex deployments across all industries with frontline workers. With Workspace ONE, organizations can quickly and easily stage, manage, and support any device—from rugged handheld computers and point-of-sale devices to wearables and self-service kiosks—alongside existing mobile and laptop deployments from a single console.

CUSTOMER PAIN POINTS

- Enrollment and configuration
- Device downtime
- Frontline worker EX

Key Differentiators

- Only UEM consistently recognized as a leader by industry analysts
- Single platform built to support any device type and use case with integrated management, identity, intelligence, and remote support
- Support for nontraditional endpoints like head-mounted wearables, peripherals, mobile printers, and interactive kiosks
- Save time and resources with low-touch enrollment and configuration, including support for side load staging, device manufacturer-specific barcode enrolment, and Android Enterprise
- Scale to support the most complex deployments with multitenant architecture and support for relay servers
- Make sure devices are always up-to-date with product provisioning
- Customize device UI, lock into single or multi-app mode, and enable check-in/check-out with Workspace ONE Launcher

- Manage the full device and app lifecycle with zero trust security and extensive policy support

- Leverage advanced device and app analytics to make data-driven decisions, like predicting battery failure and automating replacement, with Workspace ONE Intelligence

- Remotely support frontline workers with device and app tasks and issues with Workspace ONE Assist

Frontline Workers Deliver Essential Goods and Services

Service and task workers have different requirements than desk-based employees.

From delivery drivers to warehouse workers, store associates, and nurses, frontline workers rely on mission-critical technologies to complete the task or operation at hand. Businesses with frontline workers will continue to invest in mobility to optimize efficiency and transform workflows.

- Frontline workers will account for 60% of US workforce by 2024.

- In 2020, only 49% of frontline workers were mobile enabled.

- 70% of new mobile investments over the next 5 years will be for frontline workers.

Mission-Critical Devices Are a Crucial Part of the Supply Chain

Frontline workers on-site and in the field have unique mobility requirements.

- Rugged mobile computers and ruggedized consumer devices

- AR, MR, and VR head-mounted wearables

- Mobile printers

- Point-of-sale or point-of-care devices, interactive kiosks, and digital signage

Mission-Critical Device Deployments in Retail

Organizations have to optimize frontline worker productivity from the back to front of store and enable store associates to easily communicate with each other, access product and customer information, and process payments with rugged mobile computers or barcode scanners and consumer devices in enterprise sleds. They have to provide workers with immersive training with VR headsets and print barcodes, labels, and receipts with mobile printers by promoting new products and create purchase impulse with interactive kiosks.

- Ensure worker and customer health by enabling cashierless payment options via beacons or cameras.

- Manage unmanned IoT endpoints, connected to your network, like sensors, security cameras and more.

- Virtualize your POS to track consumer identities and habits and empower shoppers with self-checkout and loyalty programs .

- Quickly scale to meet market demand and ensure store associate safety by allowing them to bring their own device to work and securely access necessary info.

- Check in-store associate and customers' temperatures before entering the store and give managers access to heat mapping technology to mitigate health risk.

- Ensure worker and customer health and decrease shipping costs by enabling curbside pickup.

- Boost profitability and protect your customers by offering online purchasing and delivery options that can be easily tracked throughout the supply chain.

- Turn your store into a fulfillment center to improve store offerings, shorten delivery windows, and safely scale shipping to meet local market demands.

- Leverage AR or VR technology to provide your distributed workforce with at-home immersive training, or to help shoppers visualize how products will fit or look before they purchase online.

- Scale to support your distributed workforce by giving them access to the apps and data they need with digital workspace technologies, including virtual desktops and apps and BYO.

- Empower and ensure the safety of the retail workforce and shoppers with VMware end-user computing (EUC).

Mission-Critical Device Deployments in Retail

We are maintaining business continuity during and after global pandemic by empowering and ensuring the safety of the retail workforce and shoppers with VMware end-user computing (EUC) solution.

- **Rugged mobile computers or barcode scanners and consumer devices in enterprise sleds:** Enable store associates to easily communicate with each other, access product and customer information, and process payments.

- **Head-mounted displays (HMDs):** Immersive, hands-on training.

- **Mobile printers:** Print barcodes, labels and receipts .

- **Interactive kiosks and digital signage:** Promote new products, create purchase impulse, and collect and access insightful in-store data.

Mission-Critical Device Deployments in Healthcare

Healthcare organizations have to optimize efficiency and reduce error rate at the point-of-care by adopting the following methodologies:

- **Rugged mobile computers or barcode scanners and consumer devices in enterprise sleds:** Collect key info and provide real-time access to patient vitals, diagnostics, imaging and more at the point-of-care.

- **Head-mounted displays (HMDs):** Reduce medical error within the surgery room and while interacting with patients.

- **Mobile printers:** Label specimens and samples at the point of collection, increasing efficiency and reducing error.

- **Interactive kiosks and digital signage:** Implement self-check-in systems can save administrative and reception staff time and improve patient privacy and wait times.

This will help in collecting key info and provide real-time access to patient vitals, diagnostics, imaging, and more at the point-of-care with rugged mobile computers or barcode scanners and consumer devices in enterprise sleds and also in reducing medical error within the surgery room and while interacting with patients with AR head-mounted wearables by increasing efficiency and reduce error by labeling specimens and samples at the point of collection with mobile printers. We can also save admin staff time and improve patient privacy and wait times with interactive kiosks.

Mission-Critical Device Deployments in Logistics

Organizations have to optimize efficiency and visibility and quickly adapt to evolving supply chain needs by applying the following techniques:

- Track the current status and location of all assets for greater productivity and cycle count efficiency with rugged mobile computers or barcode scanners.

- Print barcodes, labels, receipts, or tickets to quickly sort materials and increase visibility with mobile printers.

- Deliver instructions, visual diagrams, and reference materials directly to workers' line-of-sight with AR head-mounted wearables.

Mission-Critical Device Deployment Challenges

Mission-critical devices have unique management requirements like the following:

- **Enrollment and Configuration:** Devices are deployed outside of the office away from IT with limited connectivity.

- **Device Downtime:** Device or app failure can cost millions of dollars a year due to decreased worker productivity.

- **Employee Experience (EX):** High frontline worker turnover due to employee disengagement is expensive.

When it comes to mission-critical device deployments, top IT challenges include:

> **Enrollment and configuration:** Since mission-critical devices are almost always deployed outside of the corporate office and away from IT, organizations must ensure enrollment and configuration be low-touch. Well-defined workflows of frontline workers (specific to their role, industry, department, task and responsibility) require extensive device and app configurations. Frontline workers in the field and essential businesses located in remote areas (or those with a high volume of shared devices, connected to a single network) also often lack basic connectivity requirements (like limited wireless coverage and data, latency issues, and low network bandwidth). This presents challenges for IT when it comes to successfully deploying business-critical security and software updates to devices. IT must also ensure data security by configuring devices to prevent unauthorized access.

> **Downtime:** Technological problems are inevitable, and when a mission-critical device or app (essential to business operations) fails or is interrupted, the worker relying on it can't do their job (i.e., complete their task or set of tasks), immediately impacting their productivity. This downtime can cost essential businesses millions of dollars a year due to decreased employee production or output and potential lost sales and customer dissatisfaction.

Employee experience (EX): While the average annual turnover rate sits at 19%, frontline worker occupations typically have a much higher turnover, with some industries (like retail and healthcare) experiencing 50–100% turnover. Employee disengagement is the number-one cause of turnover and the biggest culprit of employee disengagement is technology (i.e., insufficient technology). Turnover is expensive and with frontline workers in record demand, essential businesses must develop strategies to retain their existing employees.

1. **Optimize efficiency and transform workflows:** Simplify management and support of mission-critical device deployments with low-touch enrollment and configuration, shared device management, device and app analytics, and remote worker support.

2. **Improve frontline worker EX:** Deliver a seamless end-user experience by only giving workers access to the apps, content, and settings they need to be productive and stay engaged across shared devices to mitigate worker disengagement and turnover. Deliver a seamless end-user experience to keep workers productive and engaged.

3. **Minimize device downtime:** Maximize productivity and EX by equipping IT and help desk staff with real-time remote support to quickly assist workers with device tasks and issues before it impacts your bottom line. Assist workers with device tasks and issues before it impacts your bottom line

4. **Scale to support any use case:** Support new
    and innovative mobile technologies that
    improve worker productivity and enable
    exceptional customer experiences, at scale, like
    BYO programs, augmented and virtual reality
    head-mounted wearables, peripherals, mobile
    printers, and interactive kiosks. Support new
    technologies that improve productivity and CX
    like BYO programs and IoT

Deliver Solutions Optimized to Manage Mission-Critical Technologies

Workspace ONE unified endpoint management platform addresses all
these challenges, enabling simplified management across:

- Any endpoint

- Any platform

- And for any use case

These features are under a single pane of glass for maximum visibility
and security.

Why a User Experience Manager (UEM)?

UEM combines MDM and EMM capabilities to create a holistic
management framework that enables organizations to manage any endpoint
across a single platform for maximized visibility and security. A comprehensive
UEM platform like Workspace ONE is especially crucial for mission-critical
deployments. IT is usually relying on too many solutions across use cases so
Workspace ONE can help them to minimize point products.

- Cut costs and eliminate multiple licenses and
  management silos with Workspace ONE.

- Consolidate management silos across all use cases with Workspace ONE, a single platform for integrated management, identity, analytics, and remote support.

- According to IDC, 70% of organizations are using multiple management tools across use cases and want to consolidate functions under a single platform.

Workspace ONE for Ruggedized Mission-Critical Devices

Organizations have to meet the unique requirements of frontline workers.

Low-touch enrollment: Save time and resources with zero-touch enrollment and configuration.

Shared device management: Customize device UI, lock into single or multi-app mode, and enable check-in/check-out to improve security and EX.

Complete lifecycle management: Securely manage the full device and app lifecycle with extensive policy support and analytics and automation.

Remote worker support: Remotely support workers in the field with device and app tasks and issues to decrease downtime.

Workspace ONE for Head-Mounted Wearables

Organizations have to meet the unique requirements of frontline workers.

Streamlined onboarding: Save time and resources with low-touch enrollment and configuration.

Simplified app deployment: Give workers access to the augmented and virtual reality apps that drive business operations.

Complete lifecycle management: Securely manage the full device lifecycle with extensive policy support across OEMs.

Remote worker support: Remotely support workers in the field with device and app tasks and issues.

Workspace ONE for personal devices: Amid the global pandemic, essential businesses have started allowing BYO to deliver critical info to workers on-site, in the field, and at home.

Workspace ONE Intelligent Hub

Workspace ONE Intelligent Hub is a unified app and endpoint management agent from VMware that acts as the user-facing component of the Workspace ONE platform. It provides a single interface for users to securely access work apps, resources, and services on any device—while also enabling IT to manage and secure endpoints.

- Enable workers to use unmanaged personal devices to access corporate apps and info relevant to their role, while maintaining employee privacy.

- Deliver personalized notifications to workers without a corporate email address, like system outage or inclement weather alerts, safety procedure updates, or shift changes.

- Keep new hires engaged and prepare them for their first day with pre-hire training and HR tasks.

Employees can use their personal, unmanaged devices for access to company apps and information while maintaining employee privacy. Employee can access any web, native, virtual, or SaaS app with SSO so they are connected and productive.

Leverage employee onboarding to keep new hires engaged and get them ready for their first day by providing access on their personal device. Employees can select the tools they need for work, complete pre-hire training, and finish any required tasks for the first day right within Intelligent Hub. For employees who might not have corporate email addresses, leverage Intelligent Hub "For You" notifications to alert employees of any company, shift, or environmental updates. It could be a system outage, inclement weather, or a request to change shifts. Give access to critical documents and intranet sites right within Intelligent Hub so employees can very easily find knowledge base articles, manuals, HR or benefits information, and anything else important to their job or work environment.

VMware is a recognized leader in retail IT transformation. A full 10 out of the top 10 Global Retailers use VMware, based on the latest RIS News report of the top 100 retailers.

Mission-Critical Deployment Trends

Consumerization of Devices

- Proliferation of use cases across frontline worker roles and tasks

- Increasing use of consumer-like rugged devices and ruggedized consumer devices

Rise of Android Enterprise

- Deprecation of legacy Android and Windows CE management

- Standardized management, advanced APIs and security with Android Enterprise

Rise of IoT and BYO

- Introduction of IoT endpoints, including wearables, kiosks, and gateways and edge systems

- Introduction of BYO to allow workers to access corporate content on-site or at home

Deprecation of legacy Android management (DA mode) in Android 10 and deprecation of legacy Windows CE by 2021.

The use of consumer devices in rugged deployment scenarios has proven that the choice of consumer devices over rugged devices for use in rugged deployments doesn't make sense—for many reasons.

#1. The initial purchase price doesn't reflect the TCO.
Unquestionably, you can purchase a consumer smartphone or tablet for less than the cost of a purpose-built, rugged version. The replacement, however, is considerably higher than the initial one. Also factor in that the consumer device has a shorter lifecycle—about two years—while a rugged device is expected to last four years or longer. Recent research by the Aberdeen Group showed that a business with 1,000 mobile devices spends approximately $170,000 more per year to support consumer-grade devices than the enterprise-grade. Another study conducted by VDC Research Group stated that consumer devices are three times more likely to fail in their first year and 77% of those cases are due to a dropped device.

#2. Consumer devices aren't equipped for enterprise needs.
Rugged devices include greater screen resiliency to light and moisture, better acoustics, and tougher drop-resistance. While consumer devices may be inexpensive and compact, a scanning application can drain the battery in a matter of a two to four hours. However, an enterprise device uses a replaceable lithium ion battery that is specifically designed for full-shift scanning and can last up to 12 hours in production.

Keep in mind that you could have a whole fleet of backup consumer devices, but you would need three times as many to make it through a whole shift. With an enterprise device, you can swap out the removable battery without disrupting the workflow when the battery runs low.

#3. Consumer devices strain your IT resources.
Mobile devices are accessing your network—and other users of those mobile devices (e.g., kids). That means your enterprise is susceptible to viruses and unauthorized access. In addition, mobile devices are easily lost or stolen. Between provisioning, remotely wiping, and upgrading mobile devices and applications on a wide range of consumer devices, you're pushing the limits of your IT staff's time. An enterprise-grade, rugged device is more easily managed through your MDM.

While consumer devices used to appear much sleeker and appealing than their rugged cousins, rugged devices have evolved to incorporate many of the features offered by consumer devices, but with more brains and brawn.

Deprecation of Both Windows Rugged and Legacy Android

Device Admin API Deprecation

- Android 2.2 (Froyo) was the original version that supported management of mobile devices via device admin APIs.

- Device admin APIs provide device administration features at the system level, such as password polices, device wipe, or camera restrictions.

- Device admin APIs became the basis for many MDM solutions to manage enterprise devices.

These are some of the challenges of device admin APIs:

- There is no ability to separate personal data on BYOD or mixed-use devices.

- There is public app management and distribution that relies on a Google account.

- In order to download applications from Google Play, you have to ensure that every user has a Google account, but then once it's present on the device, users can download any applications they want, back up data to Google's servers, or accidentally lock devices with Android's Factory Reset Protection (FRP).

- There is inconsistent management across devices from different OEMs.

- It is an all-or-nothing approach; any app can be designate as a device admin.

Rise of Android Enterprise: A Modern Management Technique

Problem Statement: Employee Experience

- In fact, 85% of workers are disengaged, costing companies millions of dollars a year in productivity loss.

- A key culprit in employee disengagement is technology, which is at the core of how today's employees stay productive and innovate.

IT Challenges Across Worker Type

According to the Help Desk Institute, 54% of organizations today are experiencing a rise in support ticket volume due to new customer equipment, devices, and apps.

> **Help desks** are using traditional, stand-alone IT service management (ITSM) tools that don't give them the flexibility they need. Outdated ITSM tools and processes, the increase in overall employee devices and apps, and high worker turnover rate are causing an overwhelming increase in ticket volume and longer resolution times.

> **Task workers** (like in-store sales associates, field workers, warehouse workers, truck drivers, etc.) rely on corporate-owned devices and mission-critical apps to get their job done. If an app is down, this could have a serious impact on a company's bottom line.

> **Knowledge workers** (like accounting and finance managers, sales managers, marketing research and business analysts, executives, etc.) rely heavily on the Internet, email, documents, presentations, and spreadsheets. They often require remote access, and mobility is frequently their number one concern. If knowledge workers are unsure of how to perform tasks on their corporate-owned tablets or laptops or have issues accessing the corporate network or data on BYO devices, it can negatively impact employee satisfaction, productivity, and inevitably, the customer experience.

One recent study showed that one dropped connection or poorly performing application per shift can cost companies as much as $20,000 per year in support and productivity losses per mobile worker. The same study showed that the consequences of every failure incident can cost companies in up to 100 minutes of lost productivity.

Workspace ONE for Mission-Critical Devices

We have to meet the unique management requirements of mission-critical deployments.

Stage

- Select the best onboarding method for your deployment and enroll devices in Workspace ONE.

- Configure Workspace ONE relay servers to distribute content to devices without compromising Workspace ONE server bandwidth and data usage.

Enroll Devices

- Android Enterprise

- Samsung Knox Mobile Enrollment

- Device Manufacturer-Specific Barcode Enrollment

Configure Relay Servers

- Pull relay servers

- Relay server cloud connector

Simplify Setup with a Variety of Enrollment Options - Seven Low-Touch Ways to Enroll Your Devices in Workspace ONE.

7 Low-Touch (or Zero-Touch) Device Enrollment Methods in VMware Workspace ONE, designed to automate onboarding and reduce IT effort while enhancing user experience:

> Apple ADE
>
> Android Zero-Touch
>
> Windows AutoPilot
>
> QR Code
>
> Hub-Based Enrollment
>
> OEMConfig
>
> Drop Ship Provisioning

Workspace ONE Rugged Enrollment Configuration Wizard

Simplify Setup with Android Enterprise Enrollment Options

- EMM Token: Easily enroll small device deployments for testing

- Side: Load and barcode

NFC Bump: Transfer configurations to new devices with NFC bump, **QR Code:** Scan QR code to enroll a device from setup wizard, **Zebra StageNow Barcode Enrollment:** Easily Stage Devices in the Workspace ONE Rugged Enrollment Configuration Wizard

Zebra StageNow Barcode Enrollment Integration

1. Zebra StageNow integrates barcode enrollment.

2. Workspace ONE sends staging info to Zebra StageNow.

3. Zebra StageNow sends barcode to Workspace ONE.

4. Using the Zebra StageNow app, scan barcode with device.

5. The device downloads staging info from the relay server and is now enrolled and configured.

Zebra StageNow is fully integrated with the Workspace ONE console and creates a staging package, generates a Zebra StageNow barcode, and scans to enroll.

Figure 2-1. *Enrollment process*

Honeywell Enterprise Provisioner Barcode Enrollment Integration

1. In Workspace ONE, create a staging package, containing device configurations, settings, and apps, and generate a Honeywell Enterprise Provisioner barcode.

2. Workspace ONE sends the staging info to Honeywell Enterprise Provisioner.

47

3. The Honeywell Enterprise Provisioner sends barcode to Workspace ONE.

4. Using the Honeywell Enterprise Provisioner app, scan the barcode with the device.

5. The device downloads the staging info from the relay server and is now enrolled and configured.

Optimize Content Distribution with Workspace ONE Relay Servers

Relay servers act as a content distribution node that provides help in bandwidth and data use control. Relay servers act as a proxy between the Workspace ONE UEM server and the rugged device for product provisioning. The relay server acts as an FTP/Explicit FTPS/SFTP server that distributes products to the device for download and installation. You can distribute to all devices without consuming all the bandwidth to the main/central MDM server.

There are two types of relay servers:

Push relay servers: This method is typically used in on-premises deployments. The UEM console pushes content and applications contained in the product or staging to the relay server.

Pull relay servers: This method is typically used in SaaS deployments. A web-based application stored in the relay server pulls content and applications contained in the product or staging from the UEM console through an outbound connection.

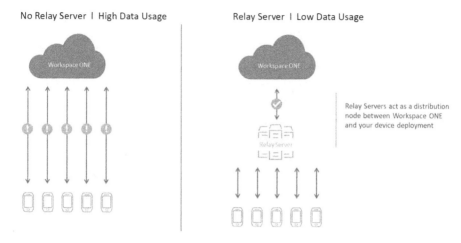

Figure 2-2. *Relay servers*

Workspace ONE Pull Relay Servers for SaaS Deployments -
Distribute Content to Devices Without Compromising Workspace ONE
Server Bandwidth

1. In Workspace ONE, create and schedule content
   and configurations to deploy to devices and choose
   to distribute via relay servers.

2. The relay server pulls new or updated content and
   apps from Workspace ONE.

3. The device pulls new or updated content and apps
   from relay server.

Workspace ONE also supports push relay servers for on-prem
deployments.

Workspace ONE Relay Server Cloud Connector - Securely Distribute Content to Devices on Hosted SaaS Environments

1. Enables single outbound (pull) connection from customer's network to the Workspace ONE

2. Pushes distribution to network of FTP relay servers for enhanced throughput of content

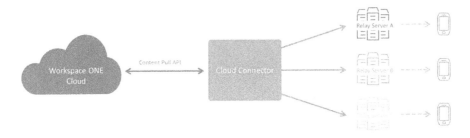

Figure 2-3. *SaaS connectivity*

Zero-Touch: Enroll devices in bulk with full management and no manual setup

Samsung Zero-Touch Knox Mobile Enrollment: Quickly Bulk Enroll Samsung Android Devices to Workspace ONE

Figure 2-4. *Honeywell bar code scanning*

Knox Mobile Enrollment is part of Knox Suite. Add thousands of Samsung devices to your enterprise at once, without having to manually enroll each one. Your users just power on their devices and connect to the network to enroll to your enterprise mobility management (MDM/EMM) provider.

Figure 2-5. *Mobile interface*

Knox Mobile Enrollment is a zero-touch deployment service that allows you to quickly enroll large number of Android devices to your MDM/EMM for corporate use. Once an IT admin registers a device with the service, the device user simply has to turn it on and connect to Wi-Fi or 3G/4G during the initial device setup process. There's no need for IMEI management and verification, and participating Knox Deployment Program resellers register your purchased devices on your behalf.

See https://www.samsungknox.com/en/solutions/it-solutions/ knox-mobile-enrollment.

Figure 2-6. *Device registration*

Once device is registered, simply turn on and connect to the network to enroll device into Workspace ONE.

Enroll Head-Mounted Wearables with Android Enterprise

- Android Enterprise Work Managed Device Mode

Figure 2-7. *Enrolling*

- Silently deploy profiles, Wi-Fi configuration, credentials
  and certificates, configure kiosk mode and VPN, and
  set restrictions

Figure 2-8. *Registration*

Android Enterprise QR Code Enrollment for Wearables

With a QR code generated from the WS1 UEM console, admins can connect a device to a staging network, indicate which version of the Intelligent Hub to download, and auto-enroll the device into an organizational group. Once the device is enrolled, configuration profiles and predefined applications can be installed onto your Realwear HMT-1 automatically over the air. And yes, in the case of Realwear headsets, this can also include WS1 Assist for remote troubleshooting of the device in the field.

Figure 2-9. *Simply scan the QR code to enroll a device from setup wizard*

Manage

- Configure devices and ensure they're always up-to-date with product provisioning.

- Customize user experience and enable check-in/check-out for multiple users with Workspace ONE Launcher.

54

- Control and monitor your deployment with custom attributes and REST APIs.

- Manage devices and content from console dashboard and make data-driven decisions across your deployment with Workspace ONE Intelligence.

Configure Devices

- Product provisioning

- Profiles

- Workspace ONE Launcher

- App lifecycle management

- Android Enterprise OEM config

- Files/actions

- Event/actions

Control and Monitor Deployment

- Custom attributes

- REST APIs

- Console dashboard

Gain Insights Across Deployment

- Workspace ONE Intelligence

Configure and Silently Install Products Over-the-Air - Ensure Your Mission-Critical Devices are Always Up-to-Date with Product Provisioning

Create products containing profiles, apps, files/actions and event/actions.

> **Conditions:** Define when product is downloaded and installed onto device.

Deployment: Control date and time products are activated and deactivated.

Dependencies: Set product to only provision devices that have certain products on it.

Figure 2-10. *Profile management*

Product provisioning enables us to create, through Workspace ONE UEM, products containing profiles, applications, files/actions, and event actions (depending on the platform we use). These products follow a set of rules, schedules, and dependencies as guidelines for ensuring our devices remain up-to-date with the content they need.

On rugged devices, we can use product provisioning, which allows us to create products that ensure devices remain up-to-date with the profiles and file/actions we assign to them. The main feature of the product provisioning system is creating an ordered installation of profiles, applications, and/or files/actions (depending on the platform used) into one product to be pushed to devices based on the conditions we create.

To deploy our products, assign them to smart groups and create dependencies. To set the product to only provision devices that have other products provisioned, set conditions, the tool that determines when a product is downloaded as well as when it is installed, product sets, which allow us to group conflicting products and rank the products based on business needs, and persistence, which allows us to enable profiles, files/actions and applications to remain on a device following an enterprise reset.

Product provisioning also encompasses the use of relay servers. These servers are FTPs servers designed to work as a go-between for devices and the UEM console. Create these servers for each store or warehouse to store product content for distribution to our devices. Another product provisioning feature is the staging methods of enrollment. Depending on the device type, we can perform device staging that quickly enrolls a device and downloads the Workspace ONE Intelligent Hub, Wi-Fi profile, and any other important content. The methods of staging a device vary by platform.

Workspace ONE Product Provisioning Components

Create Products Containing Profiles

- Create profiles that are installed or uninstalled as part of a product provisioning.

- Control device Wi-Fi, passcode, and Workspace ONE Launcher settings, restrictions, or VPN connections.

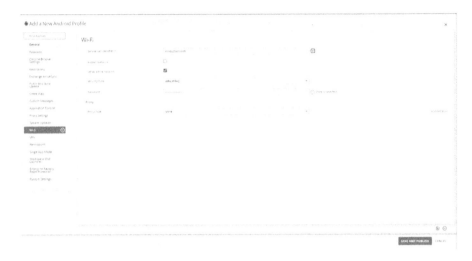

Figure 2-11. *Adding a new profile*

Workspace ONE Launcher

- Customize and lock shared or unmanned devices into kiosk mode.

- Configure in single, multi-app, or template mode.

- Customize app access and background.

- Enable check-in/check-out.

Figure 2-12. *Workspace Launcher*

For organizational use cases such as kiosk and user-less devices, AirWatch provides IT with options to easily configure in a single or multi-app mode and prevent user access to settings with VMware AirWatch Launcher. Customize application access, device background, size, and configuration of icons, and specify available settings. For example, configure a retail device into a product lookup kiosk or lock down a field device to the two data input apps required for the job.

With Workspace ONE Launcher, customize and lock shared or unmanned devices into single, multi-app, or template mode. Workspace ONE Launcher also improves the security of shared devices and boosts shift worker EX, with check-in/check-out. With check-in/check-out, assign Launcher profiles based on worker role, so employees are only given access to the settings and apps they need to do their job.

Since deploying a device to each user can be cost-prohibitive and often employees leveraging ruggedized devices are working in shifts, many organizations configure a single device to be shared by multiple users. AirWatch enables a user to authenticate into a device, or check out, and then dynamically deploy the specific apps and settings for that user to the device. Once the employee is done using the device at the end of their shift, they simply check in the device, and it's back to a blank state and ready for the next user. Settings can be configured across an organizational group or specific to individual users. Shared devices remain enrolled and under management during the check-out and check-in process, so devices remain secure, even when not in use.

Workspace ONE Launcher Check-In/Check-Out

Enable Multi-user Devices for Retail

- Create and assign Launcher profiles, based on worker role or individual.

- Workers check out shared devices via SSO.

- Devices are configured with the access, settings, and apps tied to that worker.

Improve Security of Shared Devices and Boost Shift Worker EX

- Once device is checked-in, device settings for that worker are wiped and device is ready to be checked-out again.

- Devices that are checked-in/not being used are still enrolled and can be managed from the console.

Enable Multi-User Devices for Healthcare

- Create and assign Launcher profiles, based on worker role or individual.

- Workers check out shared devices via SSO.

- Devices are configured with the access, settings, and apps tied to that worker.

Enable Multi-User Devices for Manufacturing and Logistics

- Create and assign Launcher profiles, based on worker role or individual.

- Workers check-out shared devices via SSO.

- Devices are configured with the access, settings, and apps tied to that worker.

Workspace ONE Launcher Speed Limit Lock Feature - Prevent Unsafe Device Use While Frontline Workers Are Driving

- Set speed at which to lock device.

- Set lock and unlock sensitivity.

- Customize message alerts to users.

Figure 2-13. Limiting speed

Support

- Remotely support workers with device tasks and issues with Workspace ONE Assist.

Built to Meet the Unique Management Requirements of Mission-Critical Deployments

Workspace ONE Product Provisioning Components

- Create products containing apps
- Give workers access to the mission-critical apps that drive business operations
- Silently install internal or public apps
- Configure settings for one or across a group
- Apply security policies and set app allow and deny lists
- View installed apps, app versions and status
- Can push apps through PP and Android Enterprise for managed work

Figure 2-14. *Push apps through PP and Android Enterprise for managed work*

Same-Day Feature Delivery with Android Enterprise OEM Config

Organizations must support all OEM APIs, so there is no need for additional development.

- The device manufacturer uploads app and options to Play Store.

- IT selects app and options in Workspace ONE.

- Play Store sends app and options to device via force NFC.

- The device receives the app and options and is configured.

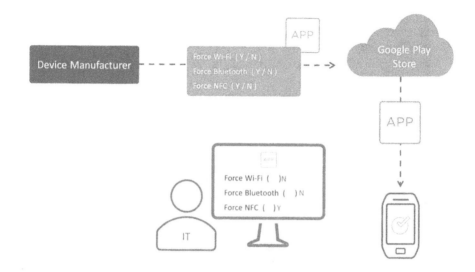

Figure 2-15. *Universal configuration for all managed devices*

Create Products Containing Files/Actions

- Files/actions are the combination of files needed on your devices and the actions you want the files to perform.

- Create an action by adding a manifest to your product with instructions.

Figure 2-16. *Actions*

Create Products Containing Files/Actions for OS and Security Updates

1. Easily push OS and security updates to your mission-critical devices with files/actions.

2. Show pushing an OS zip file to device (image), forwarding a file to a device.

3. Capability to update OS on mission-critical devices because all OEMs handle it differently.

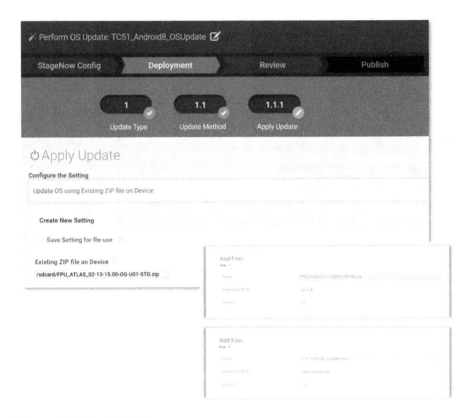

Figure 2-17. *OS Update*

Create Products Containing Files/Actions for Zebra StageNow

XML - Utilize Zebra StageNow functionality to import and export XML files to and from devices

Figure 2-18. *Adding files/actions*

We can utilize Zebra StageNow functionality to import and export XML files to and from devices.

Create Products Containing Event/Actions

Easily automate device actions based on conditions set in the console based on device-side IFTT.

IF THIS... (EVENT)

Memory Threshold

Network Threshold

Schedule

Charging - *New!*

...THEN THAT (ACTION)

Send/Receive Files

67

Launch App
Terminate App
Reboot
Check-In/Out – *New!*

Problem:

- Associates forget to check in devices when ending their shift.

- Maintenance or updates are needed at the end of a shift or when a device is unused.

Impact:

- Devices potentially left logged in, exposing data

- Users potentially completing tasks as the previous user of the device

- Increased automation complexity, time-based triggers, or manual maintenance

Event/Actions Top Use Cases

- Automatically check in device when returned to charging cradle to ensure security of multi-user devices

- Only push device and app updates when device is not in use or on network

- Only collect log files when device is returned to charging cradle

Create Custom Attributes to Control Products

- Provision products to devices based on custom attributes and their values.

- Extract specific values from devices and send back to Workspace ONE.

- Assign attribute values to devices for use in product provisioning or device look-up.

- Control assignment of products to devices that meet designated criteria .

- Create reports, dashboards and automation based on custom attributes, within Workspace ONE Intelligence.

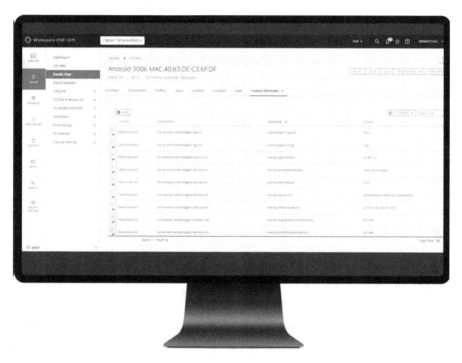

Figure 2-19. *Details view in console*

- Can use CA in filters when pushing products to devices and report on them in Intelligence.

- Custom attributes enable you to extract specific values
  from a managed device and return it to the Workspace
  ONE UEM console. You can also assign the attribute
  value for device lookup values.

- Create a custom attribute and values to push to devices.
  These attributes and values control how product rules
  work. Custom Attributes also function as lookup values
  for certain devices.

Leveraging REST APIs for Workspace ONE Deployments

Leverage REST APIs to give non-IT personnel, without access to
Workspace ONE console, control of specific functions

Customize Workspace ONE Console Access and Integrate with Third-
Party Solutions

Quickly send REST API calls to Workspace ONE via integration with
third-party solutions

Figure 2-20. *Customizing the console*

Workspace ONE Console Dashboard: Easily manage devices and content
from a single pane of glass.

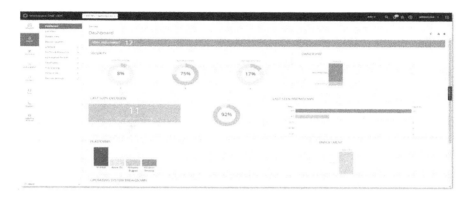

Figure 2-21. *Graphical interface*

Workspace ONE Intelligence for Mission-Critical Devices: Boost EX by proactively monitoring device health.

- Low network strength

- Unexpected shutdown

- Battery health and failures

Workspace ONE Intelligence for Mission-Critical Devices
Proactively monitor battery health, charge cycle count and failure.

Figure 2-22. *Monitoring endpoint devices*

Use battery metrics from historical data to predict failure, remediate issues, and automate replacement.

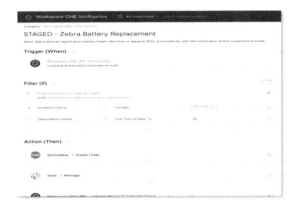

Figure 2-23. *Battery metrics*

Workspace ONE Intelligence-Based Battery Management

Use battery metrics from historical data to:

- Predict failure

- Remediate issues

- Automate replacement

Figure 2-24. *Dashboard*

72

- **Workspace ONE for Mission-Critical Devices:** Built to meet the unique management requirements of mission-critical deployments

- **Remotely Support Workers:** Workspace ONE Assist

Workspace ONE Assist Remote Support

- Decrease device downtime and maximize worker productivity and EX.

- View and control devices in real-time to quickly assist workers with device tasks and issues.

- Highlight items and guide employees through various tasks with Screen Draw.

- Notify employees when their screen is visible and enable them to pause a remote session for enhanced privacy.

- Securely access and service unattended corporate-owned devices between shifts or overnight.

- Record remote sessions for training.

- Integrate with Workspace ONE Intelligence and ServiceNow.

Figure 2-25. *Remote support console*

Workspace ONE Assist is a comprehensive remote support solution that empowers IT and help desk staff to troubleshoot and fix device problems in real-time.

With Workspace ONE Assist, IT can remotely control Android, Windows CE, and Windows 10 devices and remotely view iOS devices, directly from the Workspace ONE console.

Key Capabilities:

- Remotely connect to any enrolled device in seconds, directly from the Workspace ONE console .

- View or control devices in real time to quickly troubleshoot and fix device, network, or application issues with file, task, and application management tools, including File Manager, Command Line, and Registry Editor.

- Leverage the Screen Draw feature to highlight items on-screen for training videos or to guide employees through various tasks.

- View and export detailed device information.

- Notify employees when their personal or corporate-owned device screen is visible, and enable them to pause or end a remote session for enhanced privacy.

- Securely access and service unattended corporate-owned rugged or mobile point-of-sale (mPOS) devices between shifts or overnight.

- Easily record remote sessions for training purposes or quality assurance.

Workspace ONE Assist for Ruggedized Devices

Remotely assist workers with device and app tasks and issues.

Figure 2-26. *Main Dashboard*

Why Workspace ONE for Mission-Critical Deployments?

- The only platform consistently recognized as an industry leader by analysts

- Rugged device management

- One platform, any use case

- Ability to manage your rugged devices alongside your other endpoints for improved visibility and security across IT

- Scalability; proven deployments of 100,000+ devices

- Relay servers; 19 of the top 20 global brands in retail trust Workspace ONE to differentiate their customer experiences and grow revenue with VMware

Single pane of glass across every device and use case: Manage mission-critical endpoints alongside other device fleets for improved visibility and security.

Multitenant architecture and proven scalability: Delegate role-based access across defined segments to support complex deployments, and deliver content to devices, at scale.

Extensive API support: Give non-IT personnel, without console access, control of specific functions.

One platform for management, identity, analytics and support: Intelligence-driven platform that integrates access control, app and multi-platform endpoint management and support

Shared device management: Easily customize multi-user device UI, configure devices in single or multi-app mode, and enable multi-user devices with check-in/check-out.

Support for IoT endpoints: Easily support light-weight IoT-type endpoints, like wearables, mobile printers, peripherals, and interactive kiosks.

Case Study: West Midlands Police Streamlines Remote Support: Second Largest Police Force in UK Empowers Officers with Workspace ONE Assist

Pain Points

- Police officers on patrol frequently having to leave their post and return to the station (sometimes hours away) when device issues occur.

- Low FCR rate, with tier 1 help desk staff frequently escalating issues to tier 2 IT experts.

Use Case

- Remote view for tier 1 help desk staff and remote control for tier 2 IT experts.

- Highlight items and guide officers through various tasks with Screen Draw.

- Record remote sessions for training purposes.

The average support call is now resolved in minutes as IT can remotely control and diagnose issues on the device. This process used to take up to five hours for officers to journey into the office for troubleshooting.

- Decrease in mean time to resolution (MTTR)

- Increase in first call resolution (FCR) rate

- 1.8+ million hours saved in time spent by officers, including in-person visits for technical support

Case Study: Home Depot Takes Customer Engagement to New Levels

One of our most publicized joint customers over the years has been The Home Depot and we securely deployed more than 100,000 Zebra devices across hundreds of stores in North America together. The devices are equipped with over 18 LOB apps and give store associates quick and easy access to product, price, promotion, and stock info. They are also mPOS-enabled to accept payment anywhere in the store, improving customer experience. The Home Depot (Retail) Increases Customer Engagement by using Workspace ONE for Devices Across Hundreds of Stores and 100,000+ Zebra TC75 devices quickly deployed via side load staging and barcode enrollment. Devices configured with 18+ line-of-business apps to empower store associates with product, price, promotion, and stock info and Devices mPOS-enabled to accept payment anywhere in store, enriching customer experience.

The Home Depot Increases Customer Engagement: The Home Depot (Retail) Uses Workspace ONE for Devices Across Hundreds of Stores

- 100,000+ Zebra TC75 devices quickly deployed via side load staging and barcode enrollment

- Devices configured with 18+ line-of-business apps to empower store associates with product, price, promotion, and stock info

- Devices mPOS-enabled to accept payment anywhere in store, enriching customer experience

BENEFITS

- Scalability, centralized management

- Simplified check-in and check-out

- Increased employee productivity

- Improved customer service

Case Study: Fortune 100 Retailer Maintains Business Continuity: Delivering Essential Supplies to Millions of Americans During Recent Health Emergency

- Support millions of mission-critical devices used by essential frontline workers to communicate with each other, track assets, access product and customer info, process payments and print barcodes, label and receipts

- Empower newly remote workforce with desktop and app virtualization solutions, BYO, SSO access and Workspace ONE productivity apps

BENEFITS

- Secure digital-first infrastructure

- Scalable mission-critical device and app deployment to maintain frontline worker productivity

- Digital workspace to enable remote work and provide a great EX

Case Study: Augmedix Rehumanizes Healthcare with Workspace ONE: Augmedix Eases the Burden of Documentation so Clinicians Can Focus on Patient Care

Horizon was the reason why we were able to maintain business continuity and move to our COVID-driven urgent business continuity phase (work from home transition) so quickly and with so few issues.

Scribes pack thin clients, take them home, connect them to their router, start using the same already secured desktop they were using from the office.

How it Works:

- Augmedix provides Google Glass or Samsung Galaxy devices to healthcare provider to use at the point-of-care.

- Augmedix assistants observe clinician-patient interactions remotely.

- Assistants extract relevant details from each visit to create medical notes and provide live clinical support.

- Medical documentation is uploaded into the clinician's EHR for sign-off.

Workspace ONE Use Case

- Manages 1,000+ devices across 14 healthcare providers worldwide with Workspace ONE UEM.

- Delivers virtual desktops and apps to assistants with Horizon.

- Amid COVID-19, Horizon was the reason Augmedix was able to maintain business continuity, by quickly and seamlessly transitioning all their assistants from the office to home.

Summary

VMware solutions for frontline workers are very efficient in today's fast-paced life and data-driven to execute frontline tasks. By leveraging VMware's advanced solutions, organizations can empower their frontline workers with the tools they need to perform their jobs efficiently, securely, and effectively. This leads to improved productivity, enhanced security, and higher overall job satisfaction, ultimately driving better business outcomes.

CHAPTER 3

SDN Business Objectives

Organizations need a next-generation VDI solution for delivering virtual desktops and applications to internal and external users from any network, using traditional and mobile computing devices. Customers are looking to deliver a standardized and seamless virtual desktop and application experience to end users (internal, remote, external, and third-party) across the enterprise. Customers have identified several business and technical reasons for using a virtual desktop and application delivery infrastructure based on VMware Horizon. They have also chosen to secure their Horizon environment with VMware NSX.

Purpose and Intended Audience

This chapter provides detailed information about the NSX integration with VMware Horizon and how software-defined networking will work in virtual or cloud environment. It will also help you start and execute a project while fulfilling the business outcome.

This chapter is intended for EUC architects, network administrators, and VDI administrators who are familiar with and want to use VMware software to deploy and manage an SDDC.

© Ajit Pratap Kundan 2025
A. P. Kundan, *Intelligent Automation with End-User Computing Solutions*,
https://doi.org/10.1007/979-8-8688-1312-2_3

Organizations need a well-defined end-user computing (EUC) strategy in place. VMware Horizon implementation aims to address the business objectives that form part of this strategy. These are summarized in Table 3-1.

This table contains sample data. In real-world use, you should replace the content with information pertaining to the actual customer.

Table 3-1. *Business Objectives*

Requirement	Description
Enhance the quality of the primary Windows desktop environment	Users are waiting too long for their existing PCs to be built and configured. A faster, more dynamic method for delivering their primary workspaces is required that does not involve manually imaging physical workstations.
Windows 11 deployment	Have to identify deployment and operational challenges that might result in an extended delivery timeline for Windows 11 to the organization. VMware Horizon can be used as the primary delivery mechanism to quickly and efficiently transition users to the Windows 11 desktop platform with little or no manipulation of the existing endpoint. This would allow to accelerate the delivery of the Windows 11 desktop using a seamless, low-risk approach.
Simplify remote access	End users currently uses a VPN-based remote access solution. As part of this effort, they want to simplify and improve the remote access experience by removing additional components and security configurations that add support and management complexity, as well as device dependence for remote access users. They want to achieve the same access to virtual desktops and applications both inside and outside the enterprise network.

(continued)

Table 3-1. (*continued*)

Requirement	Description
Defined service offerings	Provision different types of desktops to users within the same user community. This is becoming unmanageable. They want to offer a service catalog-based approach for desktop and application delivery to eliminate the complexities of managing multiple different types and configurations of endpoints.
Centralized management	Current server virtualization efforts have shown the benefits of virtualization, including improved operational efficiency, enhanced availability of servers, the ability to quickly deliver IT services, and centralized management. The virtual desktop and application delivery platform provides the ability to enhance and increase operational and financial efficiencies gained by sharing and securing desktop computing resources through virtualization.
Flexible working arrangements	Users are limited to accessing resources from company-owned and managed devices. This limits their ability to provide for flexible working arrangements. VMware Horizon supports access from virtually any modern device type, allowing admin to offer more flexible working arrangements.
Zero Trust Security	Customer wishes to improve their security profile by moving to a Zero Trust Security model. In this model users will only have access to resources they have been specifically granted access to.
Just in Time Desktop Delivery	To improve desktop management, all users should use a Common Operating Environment with any additional applications and firewalls loaded based on the user's identity. This will facilitate faster delivery of operating system updates and improve overall security.

Use Case Definition Framework

Classifying users based on common attributes is essential to the organization's solution design. The following attributes were considered in defining the sets of use cases. This data was developed from both automated assessment data and assessment workshops carried out with customer. These are broad classifications. The design should have the flexibility for outlying cases where necessary. This process can be repeated when commissioning new classes of users in the future.

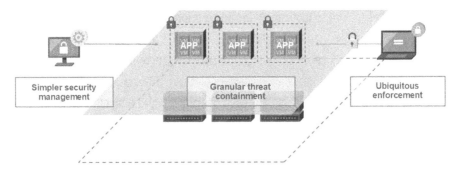

Figure 3-1. Network virtualization

Table 3-2 contains an extensive list of use case definitions. You can modify these to align with the appropriate use cases of your customer.

Table 3-2. Use Case Definition

Attribute	Definition
Number of Users	The number of users that fall into this category.
Department/Business Unit/Workgroup	Most use cases generally map to departments, business units, or workgroups, although not always.
General Use Case Function	This is a summary of the use case. For example, Contact Center.

(continued)

Table 3-2. (*continued*)

Attribute	Definition
Application Summary	While the full set of applications cannot be listed here, it is useful to understand a summary of the typical application types during the design phase. For example, light office user or intensive 3D designer. Full application usage will be available as part of the assessment reports.
Application Network Access	Like the application summary, the full list of application network flows is an important part of the use case in later stages.
Users' Application and File Home Site	Where are the users' file shares located? Where are the users' backend application servers? This information is useful for determining the appropriate data center for the client OS.
Users' Physical Location	Where is the user based? Are they always based there? Are they mobile?
Users' Network Connectivity	What is the user's network connectivity in terms of connection type, reliability, bandwidth, and latency?
Peer-to-Peer Traffic	Should the desktops be able to communicate with other desktops?
Authentication	How should these users authenticate? Does their authentication mechanism change depending on their location or what they are accessing?
Access Device	What is the planned access device for use with the new VMware Horizon platform?
Offline Requirements	Do they have a requirement for offline use? For example, on an airplane?

(continued)

Table 3-2. (*continued*)

Attribute	Definition
Time of Use	Specify the times and days when the desktop will be used. Do these users have fixed shifts?
Elevated Rights	Do these users require elevated rights? For example, installing applications.
Active Directory Domain	In which Active Directory/LDAP domain do the users (and computer accounts) reside?
Resource Requirements	In summary, what are the typical resource requirements for these users? While the assessment data will be used to accurately size the resources for these use cases, this is useful summary data that can be used for pool design.
Unified Communications	Do these users use unified communication applications, such as Skype for Business?
Multimedia Requirements	Do these users have specific multimedia requirements, such as the ability to watch high definition videos?
Flash	Do these users have extensive Flash requirements? Where is their Flash media hosted?
Web Browsing Limitations	For example, are these users ordinarily allowed to access external Web sites?
Audio Requirements	Do these users have any specific audio requirements, including microphone use?
Hardware Accelerated Graphics	Do these users require the use of hardware accelerated graphics? It is useful to group these types of users together as it affects the design.
Printing/Scanning	Do these users print? How do they currently print? Do they use specialized printers?

(*continued*)

Table 3-2. (*continued*)

Attribute	Definition
Display Specification	What is the users' display specification in terms of number of screens, resolution, and color depth?
Peripheral Devices	Do these users have a requirement for specific peripherals? For example, webcams or scanners.
Resiliency and Availability Requirements	What are the requirements in terms of RTO and RPO of the users hosted application or desktops? Individual application resiliency and availability is not in scope.

Example Use Cases

The following tables are populated with sample use case definitions. Please replace the content with information pertaining to your actual customer.

Use Case #1: Call/Contact Center

The following table describes a use case for a call/contact center:

Attribute	Description
Number of Users	500 concurrent
Department/Business Unit/Workgroup	Primary Contact Center
General Use Case Function	Contact Center
Application Summary	Outlook and CRM app only
Application Network Access	Outlook: 443 TCP to Exchange Server CRM: 80 and 443 to CRM Web Server

(*continued*)

Attribute	Description
Users' Application and File Home Site	Primary Data Center
Users' Physical Location	Primary Call Center. No mobility
Users' Network Connectivity	Well-connected LAN
Peer-to-Peer Traffic	Not Allowed
Authentication	AD credentials only
Access Device	Repurposed PC
Offline Requirements	No
Time of Use	08:00 to 18:00 weekdays
Elevated Rights	No
Active Directory Domain	abc.com
Resource Requirements	Very light office use
Unified Communications	No
Multimedia Requirements	None
Flash	No
Web Browsing Limitations	All standard
Audio Requirements	No audio. Telephony is separate
Hardware Accelerated Graphics	No
Printing/Scanning	No printing
Display Specification	1x 1280 x 1024 x 24-bit
Peripheral Devices	No
Resiliency and Availability Requirements	20 minutes RTO. No data on desktop, so RPO is a new desktop

Use Case #2: Mobile Sales Team

The following table describes a use case for a mobile sales team:

Attribute	Description
Number of Users	100 concurrent
Department/Business Unit/Workgroup	APAC Sales
General Use Case Function	Mobile Sales Team
Application Summary	In-house pricing tool
Application Network Ports	In-house pricing too: 80 and 443 TCP to Web Server
Users' Application and File Home Site	Primary data center
Users' Physical Location	Mobile
Users' Network Connectivity	4G
Peer-to-Peer Traffic	Not Allowed
Authentication	RSA SecurID
Access Device	Laptop
Offline Requirements	Yes, but local apps on laptop
Time of Use	08:00 to 18:00 weekdays
Elevated Rights	No
Active Directory Domain	xyz.com
Resource Requirements	Very light office use
Unified Communications	No
Multimedia Requirements	None
Flash	Yes

(*continued*)

91

Attribute	Description
Web Browsing Limitations	All standard
Audio Requirements	Output only for training videos
Hardware Accelerated Graphics	No
Printing/Scanning	Local USB printer
Display Specification	1x 1280 x 1024 x 24-bit
Peripheral Devices	No
Resiliency and Availability Requirements	Variable, return to base for new laptop. No data on desktop so RPO is a new desktop.

Nonfunctional and Technical System Requirements

This section describes the nonfunctional and technical requirements that must be met by organization's VMware Horizon design. These are requirements that must be met by the system as a whole and are not specific to individual use cases.

Table 3-3 is populated with sample data. Please replace the content with information pertaining to the actual customer.

Table 3-3. Nonfunctional and Technical Requirements

Category	Description
Printing and Peripherals	The virtual desktop and application solution must provide mechanisms to control printing and peripheral device features and functionality. USB device control and printing capabilities can be enforced by policies, Active Directory Group Policy, and USB device filtering through the endpoint/desktop.
Access and Authentication	VMware Horizon can integrate with their current authentication solutions (directory services and two-factor authentication mechanisms) for internal and external access. Desktop and application access and provisioning should be managed through current authentication mechanisms.
Systems Monitoring and Management	VMware Horizon solution to integrate with their current monitoring solutions. Critical components, such as View Connection Server instances, View Security Server instances, and VMware ESXi hosts should be proactively monitored for system health and capacity management.
Application Integration	Software and application access must be accurately monitored and controlled to maintain license compliance.
High Availability and Resiliency	The virtual desktop and application solution must be designed with resiliency and availability in mind. VMware Horizon infrastructure components must be designed to support 100 percent of the user population if any single component is rendered inoperable.

(continued)

94

Table 3-3. *(continued)*

Category	Description
Desktop Security	Secure the delivery of desktops and applications to untrusted endpoints and 3rd party vendors/contractors.
Application Integration	VMware Horizon solution integrate with VMware ThinApp, which is the standard method of virtualized application delivery in the virtual desktop enterprise.
Availability	Availability requirements of individual use cases are highlighted on a per-use case basis. The availability requirements for the overall service was determined to be 99.9% or 8.8 hours/year.
Recoverability	Recoverability is typically measured in terms of recovery time objective (RTO), which corresponds to the time required to recover a server or environment to a functional state, and recovery point objective (RPO), which corresponds to the maximum allowable data loss for the environment. The requirements for this will vary on a per use-case and per-application basis.
Security – Authentication	The system must be capable of supporting the authentication methods identified for each use case.
Security – Certificates	The system should not use self-signed certificates.

(continued)

Table 3-3. (*continued*)

Security – Desktop	VMware Horizon must integrate and comply with desktop security standards and wants to establish desktop and user security enhancements for the virtual desktop and application infrastructure. These enhancements include additional security configurations and virtual machine network isolation using firewall or virtual machine fencing concepts and technologies.
Security – Desktop	VMware Horizon provides controls and mechanisms to prevent the transfer of data between the endpoint and the data center, thus keeping applications and sensitive information secured.
Security – Network	Virtual Desktops should not be able to communicate with other virtual desktops (Peer-to-Peer Traffic), unless otherwise specified.
Security – Network	Provide the capability to monitor the network flows of the virtual desktops.
Security – Network	Access to network resources should only be granted to users who are a member of an approved Active Directory Security Group.

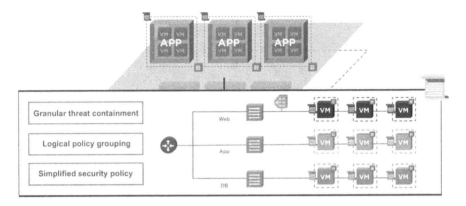

Figure 3-2. *Granular policy enforcement at VM level*

Design Constraints

Constraints can limit the design features and the implementation of the design. VMware has identified the design constraints described in Table 3-4.

This table is populated with sample data. Please replace the content with information pertaining to the actual customer.

Table 3-4. *Design Constraints*

Type	Description
Servers	The proposed CPU type and clock speed might limit the performance of users' applications and desktops.
Network Bandwidth	Current network load can affect the performance and system response of the implementation design.
Compliance	The security compliance design component can lead into a separate service offering.
Performance	For best performance, multiple machines are required for installing different components on different machines.

NSX Monitoring And Diagnostic Tools: https://docs.vmware.com/
en/VMware-NSX-Data-Center-for-vSphere/6.4/com.vmware.nsx.admin.
doc/GUID-985C019C-D65A-4DB2-A771-20437D205441.html

Design Assumptions

Assumptions are expectations about the implementation and use of a
system that cannot be confirmed during the design phase but are still
used to provide some guidance in the design. If assumptions are not met,
the respective design areas are at risk of failing to address the solution
requirements.

Table 3-5 shows design assumptions. It is populated with sample
data. Please replace the content with information pertaining to the actual
customer.

Table 3-5. *Design Assumptions*

Category	Description
Databases	Provides a dedicated database server to host VMware vCenter databases.
DNS	It is assumed that name resolution for all servers is properly configured and that both forward and reverse FQDN lookups work as expected.
Storage	Provides the correct type of storage in terms of capacity and performance for building the environment.
Network bandwidth	Provides sufficient network bandwidth to support the requirements.

(continued)

Table 3-5. (*continued*)

Category	Description
User skill	Users, specifically with administrator privileges, have sufficient knowledge of VMware products and other technologies that will be used as part of the design.
Infrastructure resources	Provides adequate infrastructure resources, including servers with adequate memory and operating systems licenses.
Product licenses	Provides all the required licenses supporting all features required in the design.
Security	All communications between the servers and desktops should be secure.

Design Risks

Risks can negatively impact the reliability of the design. VMware has identified the risks shown in Table 3-6 that could impact the design. The table is populated with sample data. Please replace the content with information pertaining to the actual customer.

Table 3-6. *Design Risks*

Category	Description
System configurations	Misconfigurations of hardware and software at the customer site can negatively impact design and project timelines.
Infrastructure	Lack of adequate infrastructure, such as limitations on network address space or lack of servers and storage, can impact the design.

(*continued*)

Table 3-6. (*continued*)

Category	Description
Architectural changes	Customer's expectation of being able to make architectural changes in the middle of the project, or adding requirements that are out of scope, will impact the design.
Hardware support	Usage of infrastructure that is not officially supported or certified can have implications during the design implementation phase.
Product licenses	The architectural design has certain expectations regarding VMware products available at the customer site, such as enterprise licenses. If these requirements are not met, the design can be affected.
Network Ports	Applications may not function correctly if accurate data on their network ports and data flows is not provided.
Missed Applications	If applications are not included in the network flows then their network access may not be granted.

External Dependencies

The delivery of the VMware Horizon project is dependent on external systems and processes. While outside the scope of this service, the project cannot succeed without them. The key external dependencies are listed in Table 3-7 so the project team can track them. The integration points with these external resources is critical and have to be planned properly.

Table 3-7. *External Dependencies*

Dependency	Description
Networking	It is important to understand the network topology when designing and implementing VMware Horizon. Elements of importance include available bandwidth (circuit speed/ type), peak and average network usage, traffic management capabilities (QoS/CoS, traffic shaping, priority queuing, and so on), and the time of day the network is most used. In addition, it is important to understand the distribution of endpoints at each site and the endpoints that are remote to the network or leverage VPN connectivity.
Directory Services and Security	Customer's core services (Active Directory, DNS, and DHCP) should be available in the data centers where VMware Horizon will be deployed. In addition, any network ports and/or firewall configurations must be considered in environments with enhanced or strict security requirements.
Change Control	This design requires deployment of production servers and Group Policy objects, which will be subject to the organization's change control procedures.
DHCP	DHCP services are required for the desktop VLANs. The availability of the DHCP servers and address space will be crucial to the availability of the virtual desktops.
Standard Operating Environment Design	Customer will provide their Standard Operating Environment (implementation of an operating system and its associated software) that will be optimized for VMware Horizon.
Application Support	Customer has existing support contracts for their applications that can be leveraged if issues are encountered.

**Configuring NSX Advanced Load Balancer for VMware
Horizon:** https://docs.vmware.com/en/VMware-NSX-Advanced-
Load-Balancer/22.1/Solutions_Guide/GUID-C7459558-1A9F-4199-
BADD-60DADE3F1D0B.html

Summary

In this chapter, you learned how to conduct workshops, customer
interviews, automated assessments, and other solution requirements
including business objectives for this digital transformation initiatives,
individual use case requirements, nonfunctional system requirements,
constraints, and assumptions along with expected risks. These solution
requirements are summarized in this chapter. We also have gone through
VMware Horizon Architecture along with NSX blueprint, which required
you to address these requirements.

CHAPTER 4

Mobility Best Practices for Higher Education

We will discuss in this chapter the admin ability to create and search by custom attributes, search for MEM devices, retrieve custom attributes for an organization group, add support for smart group tags, and retrieve details of devices where provisioning failed or is in progress. We will also discuss the device ability to search device custom attributes, results containing devices and product assignment info, compliance policies, and device security info. Finally, we will discuss the user ability to create new device custom attributes for registered devices and retrieve enrolled device details.

Purpose and Intended Audience

This chapter provides detailed information about the VMware Workspace ONE components and services including how to use them in different customer environments. It will also briefly talk about how to use the mobility solution in educational organizations.

© Ajit Pratap Kundan 2025
A. P. Kundan, *Intelligent Automation with End-User Computing Solutions*,
https://doi.org/10.1007/979-8-8688-1312-2_4

This chapter is intended for EMM administrators and VDI administrators who are familiar with and want to use VMware software to deploy and manage various kind of mobility devices.

Device Lifecycle Management with Enhanced Visibility

VMware has made several enhancements to make the IT admin experience easier for enhanced administration and serviceability and also improved device lifecycle management through these features:

- Enhanced flexibility and performance for compliance violations

- Additional notifications for enrollment workflows

- Optimized shared device and kiosk workflows

  - We've also provided enhanced visibility into system processes through the following:

    - Visibility into compliance status details/ troubleshooting

    - Visibility into the device command queue from the console

    - Additional logging for profile installation errors, smart group assignment changes, and tunnel usage

    - Helpful details for Apple APN certificate management

  - Improvements to the user interface across web app management, profile management, the end-user self-service portal, telecom dashboards, and admin role management with a new tool for role comparisons

  - Bulk management actions for users and devices

Enabling the Connected Campus

Today, laptops and mobile devices are nearly essential on a college campus. Students and teachers alike are using smartphones, tablets, e-readers, and laptops to complete and grade assignments, distribute and access educational resources, and collaborate. Many students entering college today own a mobile device or have used one in their classrooms. As a result, higher education students are more reliant than ever on mobile devices. A survey of 500 college students found that 90 percent of students surveyed say they save time studying by using mobile technology. The study also revealed that an astounding 40 percent of currently enrolled college students can't go 10 minutes without accessing some sort of technology. There are unique benefits in education that come with the rise of mobile technology. Students, with the help of tablets, laptops, smartphones, and e-readers, now have the ability to access learning materials around the clock and from any location. The in-class experience can be enhanced through technology as well, through the use of applications and other multimedia content. Professors are now able to break down the barriers of the traditional classroom with interactive take-home assignments and innovative blended learning techniques.

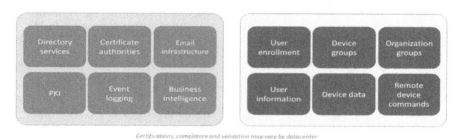

Certifications, compliance and validation may vary by datacenter

Department heads and faculty can communicate important information to students like never before. Even though bring your own device (BYOD) is not a new concept in higher education, regulations governing student information paired with increasing awareness of the threats of data loss and security breaches are moving IT departments to action. IT departments need a way to account for the devices that are accessing school networks. Without network and device security, a malicious application could potentially expose thousands of students' financial and personal information and put organizations at risk. Higher education IT departments are working to stay a step ahead. But they need comprehensive and best-in-class security solutions to do so.

Understanding Use Cases and Ownership Models

Understanding how students and professors want to use mobile devices in higher education is the starting point for a successful mobile initiative. Some may be primarily concerned with providing basic email access and a secure Wi-Fi connection, while others will want to digitize assignments and tests for completion on mobile devices. Higher education institutions should outline a mobility strategy that can adapt to the challenges of the initial deployment and enrollment but also scale to enable organizations to move more processes to mobile over time.

IT administrators should also find out who owns the devices used on campus and what the implications are for management. Many college students bring their own devices to school, and an increasing number of schools are also providing shared devices in the classroom, distributing devices, or providing a stipend for students to purchase a device.

Bring Your Own Device

A mobile device management solution with support for BYOD gives schools the ability to enable students to use their preferred devices without compromising data security. Today, students often bring personal laptops and tablets with them to school with the expectation that these devices will help them complete schoolwork. IT professionals can protect the security of their network by ensuring they have visibility into who is connecting to the network, including endpoint devices and the people using them. By leveraging the Workspace ONE solution and a network access control provider, administrators can require personally owned devices to enroll in MDM before gaining access private networks. IT administrators must also collaborate with other stakeholders to outline a clear BYOD policy that meets campus security requirements. Administrators can use Workspace ONE to distribute applications and content to devices over the air (OTA) so students can get the most up-to-date materials anytime, on the device they already own and prefer to use with a BYOD program in place. Choosing a solution that supports all device types and operating systems will ensure all students who bring their own devices will have the same access to resources.

End-User Computing Vision

While most colleges allow BYOD, many also distribute devices to their trainees for use in class for a semester or for their entire college career. Some colleges distribute tablets to incoming freshman that are preloaded with all prerequisite course materials. Some schools may purchase specialized devices that are used when students are in certain classes, such as rugged devices to be used for fieldwork in forestry and geology classes. Others provide stipends for students on financial aid or scholarship to purchase a device.

Government and defense pay for a trainee's education, and each semester trainees are given a book allowance. "What we realized after that first semester was that we needed to require them to buy the extended AppleCare," said Educational and Training institutes. Trainees are responsible for the upkeep of their device, thus taking the burden off of the Training institutes by purchasing AppleCare.

Training academies that allow BYOD and have academy-owned devices have what is called a *hybrid* deployment model. Workspace ONE enables singular management of hybrid deployments where an administrator can enroll and manage both BYOD and school-owned devices in a single administrative console. Class and other parameters enable administrators to easily push content and applications relevant to those groups to devices by grouping devices by ownership model.

Choosing Solution to Support Required Use Cases

Higher education IT administrators should look for a solution that meets their requirements now and in the future. The capabilities listed in this section will meet most organizations' needs, though administrators should weigh the importance of each one against specific and anticipated use cases. "If you can think five to 10 months ahead then you are doing pretty well," said heads of educational and training institutes. This is a good mobile management strategy. "Things aren't going to be sneaking up on you if you are thinking about them way before they come up."

Scalability

Scalability is the most basic requirement of a solution that will allow schools to plan ahead for mobility. As more devices inevitably enter college campuses, both in students' hands and as teaching tools, it will be necessary for administrators to have bandwidth in their networks and their enterprise mobility management structure so they can continue adding devices without adding strain.

Workspace ONE is designed to support deployments of any size, from dozens to tens of thousands of devices, so organizations can support an unlimited number of devices without sacrificing management capabilities.

108

Application servers are stateless by design and operate behind a network load balancer for instant and infinite horizontal scaling. These features help to reduce the up-front investment cost and allow for additional remote capacity when needed. Workspace ONE also seamlessly integrates with many of the leading network access control vendors in the market.

Device Agnosticism

A survey found that 89 percent of colleges and universities allow students to use their own device at school. True BYOD support entails supporting whichever devices and operating systems students choose to use. Choosing a solution that is device and operating system agnostic will ensure all students can use their devices for school, regardless of which device they choose or what software they're running. Moreover, it will ensure that future use cases will be accounted for, such as when a professor wants to use a previously unintroduced device type for a new class.

A Comprehensive Mobility Platform

While basic MDM and BYOD support are typically the first steps for college IT administrators, finding a platform that supports app-level management, mobile file sharing, secure browsing, secure email, and other capabilities will ensure support for further mobile integration into the learning environment.

Role-Based Administration

A solution that allows IT to delegate administration can improve app and content management by empowering trainers to distribute apps and documents directly to their students' devices. Trainers can pass out an entire semester's worth of reading material by simply dragging the assigned PDFs into a specified network folder. The trainer is then empowered to manage their students' access to certain materials, without IT intervention.

A Strong Partner Ecosystem

College IT administrators should ensure that the solution they choose integrates with other technology providers their organization uses, such as e-book publishers, hardware manufacturers, and mobile application development platforms.

Preparing Your Network Infrastructure

A College Explorer study found college students own an average of 6.9 connected devices. Even more devices will enter the fray, and data usage will rise exponentially in the coming years. Colleges must act now to ensure their wireless networks can handle the impending flood of devices and traffic. Wireless networking should not explode at any point in time and can have upward of 30% devices on our wireless a day. Universities need to manage which of those devices can access the campus network. This a constant challenge, but Workspace ONE will be able to help you.

As a best practice, it is recommended to set up a secure Wi-Fi network for students and faculty and a public network for visitors. Within the management console, administrators can limit Wi-Fi connections on managed devices to the secured network, so devices with access to student information cannot access the less secure public Wi-Fi network. Many use technology that detects new or unmanaged devices that attempt to join the secure network. Workspace ONE then redirects those devices to the Workspace ONE agent or enrollment URL to enroll (or re-enroll) a device.

There has been little emphasis at universities to lock down campus networks in the past, but in today's environment, you just can't allow that to happen. Schools have to set up both a private network and a public network. The academy is also part of a group of higher education institutions across the world that have partnered to enable traveling trainers and employees to authenticate with their academy's credentials to easily access the secure network.

Preparing Students and Professors

It is up to IT administrators with the appropriate technology to set expectations for mobile device management with students and teachers. Properly communicating the benefits and the expectations of mobile device usage on campus will help foster a secure, productive mobile environment. Communicate to students that they can choose whether or not to enroll their personally owned devices, and clearly outline what data IT will and will not be able to access. Easing students' worries about privacy will make it easier to communicate the benefits of enrolling: secure, approved access to academy resources, apps, Wi-Fi networks, e-books, other content, and more. At many colleges across the world, students are swapping heavy backpacks for school-issued iPads with electronic textbooks. Though switching to mobile technology may be second nature to young students, trainers may be less enthusiastic about allowing it in the lecture hall. Colleges should consider providing training and hosting discussions to sort out issues related to using mobile devices in class.

Designing a Smooth Enrollment Process

Institutions implementing BYOD should make it a priority to distribute clear MDM enrolment instructions via email, through video, or during a training session. Institutions can choose one or more of the following enrollment methods:

QR Code:

IT can distribute a QR code via email, which students can scan to begin enrollment.

URL:

IT can provide an enrollment URL. After navigating to the specified URL, users are prompted to enter their institutions credentials.

Agent-based:

Users can also download the Workspace ONE agent from the app store to begin enrollment.

Institutions should consider creating enrollment guides with instructions to help students and staff members enroll their own devices to minimize help-desk calls.

Leveraging the Apple Device Enrollment Program

The Apple Device Enrollment Program (DEP) provides a seamless process for enrolling Apple smartphones and tablets into MDM by integrating prompts for enrollment into initial device setup. DEP, which replaces Apple Configurator, does not require devices to be tethered for initial configuration. With DEP, the need for physical staging or provisioning processes can be nearly eliminated. DEP lets students enroll into MDM during the device's activation process. Referred to as zero-touch enrollment, this feature allows administrators to quickly enroll hundreds or thousands of devices at a time. In addition, with DEP and Workspace ONE, administrators can skip and customize steps in the device's activation process to ensure all students have the same enrollment workflow. DEP also drastically reduces the number of post-enrollment steps through the use of silent application installations. For more information, read the DEP Program Guide from Apple. Program enrollment instructions are located on the second page.

Managing Apps and Content

Let's say a book called *The History of Warfare* is an iconic textbook at a military academy and every student carries a copy with them everywhere they go, in iBook form on a school-issued iPad. In addition to this book, trainers are now writing and distributing their own textbooks in the form of iBooks, which with iOS can be managed from the Workspace ONE console. Students and professors are designing apps that replace textbooks or help students complete assignments on mobile devices at other schools. As apps and electronic content become more central to higher education, IT administrators will need a way to centrally distribute, manage, and secure them.

App Distribution and Management

Administrators can create smart groups to help with app distribution with Mobile Application Management. Smart groups enable administrators to quickly divide a mobile deployment into groups, by major, year of graduation, and other distinctions, and then distribute apps and content based on those groupings. Smart groups also enable administrators to deliver applications OTA through volume licensing programs, ensuring licenses are distributed to the appropriate students. Apple's Volume Purchase Program (VPP), Google Play for Education, and Microsoft in Education all provide ways for educational institutions to prepurchase apps in bulk for their students and staff. When a student enrolls into enterprise mobility management and downloads an app that has been prepurchased, there is no need for the student to enter any financial information. Workspace ONE can help administrators manage app purchase, licensing, and distribution in conjunction with a volume licensing program.

IT administrators can manage the entire application lifecycle with Workspace ONE, from testing to deployment to new versions to retirement. Administrators and developers can test applications in Workspace ONE with a controlled release, limiting deployment to a control group to test for issues before widely deploying the app. Controlled testing can be useful for testing a student-developed app before making it widely available. Workspace ONE offers app versioning, which enables IT administrators to require app updates, an action that can be performed organization-wide or through a phased rollout. Administrators can make apps available to users through the App Catalog, a custom app catalog where students and staff can access and download both internal and third-party apps, which can be managed with Workspace ONE.

Mobile Content Management

Sensitive files can now be distributed to student and faculty devices from an e-textbook to a grade book. Secure content lockers provide a secure and easy-to-access portal where textbooks, documents, or videos can be stored, updated, and distributed by IT or designated administrators such as professors or department heads.

Secure content locker also integrates directly to backend content repositories to provide a seamless and real-time mobile content storage system. Students can add content to their mobile devices or turn in assignments to professors' personal content lockers through the self-service portal.

Cloud Deployments

	Shared Cloud		Dedicated Cloud
Environment	Available; set up as redirects	**Custom URL**	Available; set up as CNAME record
	Admin console URL securely accessible over Internet	**Accessibility**	Admin console URL can be further locked down to customer-provided IP range
	Migration between shared cloud/ on-premise requires re-enrollment	**Migration Cloud/On-Premise**	Seamless migration between cloud/on-premise with no end user action needed
Change Management	Environment upgraded on quarterly release cycle based on change management and scheduling SLAs	**Upgrades**	Customer approval required to upgrade environment
	Testing available in shared UAT environment	**Testing**	Testing available in shared UAT environment Additional mock upgrade can be completed in dedicated environment with production backup and made available to customer to certify prior to production upgrade
	Applied on regular interval, based on change management and scheduling SLAs	**Hotfix Updates**	Applied on regular interval, customer approval required.

Data Security

Data-at-Rest Protection and Data-in-Transit Security.

Penetration and Vulnerability Testing

A comprehensive cybersecurity testing and reporting framework which typically includes internal and external testing, along with detailed documentation of attacks, defenses, and improvement actions.

Data Destruction

The DoD mandated 5220-22M for the physical destruction of data (where applicable).

Secure Content Locker

Data a -rest is encrypted with AES 256-bit encryption, data in transit is encrypted with industry- standard SSL encryption, and a secure channel certificate is used to prevent man-in-the-middle attacks.

Data Privacy and Access Control

Data Privacy and Access Control as it applies to Enterprise Mobility, ZTNA, or UEM solutions like Workspace ONE and can be used in policies, audits, or in solution architecture. Privacy is what protects people. Access control is what protects data.

Privacy

- Ensure personally identifiable information (PII) is not stored with customizable settings.

- Encrypt user authentication credentials and confidential data.

- Implement controls to ensure customer data is secure with third-party validation groups.

Access Control

- Internal team with no access
- Team separation of duties
- Principle of least privilege
- Monthly manual audit of access groups
- AD role changed and all device data wiped upon employee departure

Logins

- Separate logins for each database
- Read, write, and execute permissions
- Encrypted login and authentication credentials
- Intrusion prevention (IP)

Enterprise Integration: Integrate with back-end enterprise systems and transmit requests to critical enterprise components.

APIs: Allow external programs to use core product functionality through integration with existing IT systems and third-party applications.

Reporting & Support: Internal utility for troubleshooting customer access, granting access to shared cloud environments for support personnel and other general department-specific tools.

Support
Global operations team available 24/7/365 with additional support from account managers.

ASK
Complete knowledge base with product documentation, FAQs, and support center for creating and tracking support requests.

Announcements

Receive alerts, notifications and announcements regarding changes, incidents/events, problems, successes, and failures.

Patch and Upgrade Management

Subscribe to vendor security and bug tracking lists to ensure we upgrade network, utility, and security equipment as needed.

Operations Best Practices

Performance

Constant monitoring of server usage and capacity to ensure optimal performance at all times.

Monitoring Tools

The global network operations center (NOC) monitors server environment performance and availability 24/7/365.

Defect Tracking

Maintain detailed bug information at a granular level by release, module, personnel, and customer.

Business Continuity Plan

Weekly Nonemergency vendor bug and security announcement review

Monthly Vulnerability management

Quarterly Key infrastructure failover tests

Annual Penetration testing, backup data center failover tests, disaster recovery procedure review

Dedicated Operations Team: Global team members in data centers operations and cloud operations teams, highly qualified and trained professionals

Comprehensive background checks along with nondisclosure and confidentiality agreements

- Technical and professional certifications and memberships:
- CISSP
- CISM
- CCDA
- VCP
- ITILv3
- CAN
- A+
- ISACA
- ISSA

Introduction to Your Service Agreement

This service provides for technical support related to enterprise mobility management. The services provide the capabilities in this service description. The solution allows customers to activate, profile, and track mobile devices and usage. This service is subject to the signed EULA agreement that explicitly authorizes the sale of this service. By placing an order for the services or utilizing the services and associated software, the customer agrees to be bound by the service description. If you are entering this service description on behalf of a company or other legal entity, you represent that you have authority to bind such entity to this service description.

Enterprise Mobility Management (EMM) Service Overview

VMware will provide implementation services connected with the deployment of Workspace ONE on-premise components in the client's data centers in a disaster recovery (DR) model. The scope of the project is for implementing a replica (or scaled-down version) of the client's current production environment of Workspace ONE. This project accounts for a possible additional two meetings to your existing bundle deployment, all conducted remotely unless agreed upon by all parties.

Meeting 1: Kick-Off

This meeting is designed to cover specifics related to your SOW. The topics covered are focused on preparing for the technical installation of system components, on the scope of the features available/recommended in your SOW, and on the pre-installation expectations of a client's infrastructure/environment.

Change Management

Meeting 2: Configuration and Deployment

This meeting is designed to install the necessary technical components to deploy Workspace ONE in a DR manner within the client's infrastructure.

Service Assumptions

- Workspace ONE virtual machines have been implemented and verified to be fully functional for all components for the DR servers to be implemented as part of this SOW. Any components not validated to be working on the original servers will be out of scope of the DR servers.

- Workspace ONE will assist with the implementation of a replica of each enterprise mobility management server type (i.e., device services, console).

- VMware will assist with implementation of up to one additional gateway server of each type (i.e., Secure Email Gateway, Mobile Access Gateway [Relay/ Endpoint], Cloud Connector).

- High availability (HA) can be purchased as the associated service offering to incorporate HA into the scope of a deployment.

- Configuration of client infrastructure components (i.e., firewalls, load balancers, and SQL servers) is the responsibility of the client.

- Procurement and installation of hardware are the responsibility of the client. VMware will provide recommendations and assistance.

- Alignment of all MDM configurations and policy design with client's requirements is the responsibility of the client. VMware will provide recommendations and assistance.

- All work, documentation, and work product(s) will be conducted during local business hours and will be provided in English.

- The statement of work is deemed to be complete upon any of the following:

  - Completion of all service deliverables

  - Up to a maximum of 12 weeks after the kick-off call has occurred

  - Up to a maximum of one calendar year from purchase date, expiring after 12 months

The success of enterprise mobility will depend on its implementation and the adaptability of the workforce at all levels; it can deliver immense benefits.

- Real-time data analysis

- Better resource allocation

- Effective partner participation

- Improved employee productivity

- Real-time collaboration

VMware is the leader in enterprise-grade mobile device management, mobile application management, and mobile content management solutions, designed to simplify mobility. Customers across the world trust VMware to manage their mobile devices, including the apps and content on those devices. VMware solutions are comprehensive, built on a powerful yet easy-to-use platform. The platform was developed from the ground up to be multitenant, highly scalable, integrate with existing enterprise systems, and more, with the flexibility to be deployed on-premise or in the cloud. Its powerful yet easy-to-use platform continues to differentiate solutions in the marketplace.

- Cutting-edge software solutions for the latest Apple, Android, BlackBerry, Symbian, and Windows technologies

- Exciting opportunities to work with global customers and the world's leading OEMs, carriers, ISVs, and more

- Commitment to being better and faster at developing innovative solutions for mission-critical mobile deployments

- Experienced leadership team, which has successfully developed and implemented mobility solutions

- Fast-growing, financially stable company with global offices in North America, EMEA and APAC

Admin Console

Administrators use the Admin Console via a web browser to secure, configure, monitor, and manage their corporate device fleet. The Admin Console also typically contains the API, which allows external applications to interact with the MDM solution; this API provides layered security to restrict access both on an application level and a user level.

Device Services

Device services are the components that actively communicate with devices. It relies on this component for processing:

- Device enrollment

- Application provisioning

- Delivering device commands and receiving device data

- Hosting the Self-Service Portal, which device users can access (through a web browser) to monitor and manage their devices

SQL Database

It stores all device and environment data in a Microsoft SQL Server database. Because of the amount of data flowing in and out of the database, proper sizing of the database server is crucial to a successful deployment. Additionally, EMM (Enterprise Mobility Management) solution utilizes Microsoft SQL Reporting Services to report on data collected by the solution.

Optional Components

Secure Email Gateway offers advanced email management capabilities such as:

- Detection and remediation of rogue devices connecting to email

- Advanced controls of Mobile Mail access

- Advanced access control for administrators

- Integration with the compliance engine

- Enhanced traffic visibility through interactive email dashboards

- Certificate integration for advanced protection

- Email attachment control

Enterprises using certain types of email servers, such as Exchange or Lotus Traveler, should use the Secure Email Gateway (SEG) server in order to take advantage of these advanced email management capabilities. The SEG acts as a proxy, handling all Exchange Active Sync traffic between devices and an enterprise's existing ActiveSync endpoint.

Enterprises using Exchange, Office 365, or Google Apps for Work should not need the Secure Email Gateway server. For these email infrastructures, a different deployment model can be used that does not require a proxy server, such as Microsoft PowerShell Integration or Google password management techniques. Email attachment control functionality requires the use of the Secure Email Gateway proxy server regardless of email server type.

Cloud Messaging (AWCM)

AirWatch Cloud Messaging (AWCM) streamlines the delivery of messages and commands from the Admin Console and eliminates the need for end users to access the public Internet and procure Google IDs. AWCM also serves as a comprehensive substitute for Google Cloud Messaging (GCM) for Android devices. AWCM is the only option to provide MDM capabilities for Windows Rugged and Symbian devices. It is typically installed on the Device Services server.

AWCM simplifies device management by:

- Removing the need for third-party IDs.

- Delivering AirWatch Console commands directly to Android, Symbian, and Windows Rugged devices

- Enabling the ability for remote control and file management on Android SAFE and Windows Rugged devices

- Reducing security concerns by eliminating device communication to public endpoints outside of AirWatch

- Increasing functionality of internal Wi-Fi only devices

Cloud Connector

Cloud Connector provides organizations with the ability to integrate Workspace ONE with their back-end enterprise systems. Cloud Connector runs in the internal network, acting as a proxy that securely transmits requests from workspace to the organization's critical enterprise infrastructure components. This allows organizations to leverage the benefits of MDM, running in any configuration, together with those of their existing LDAP, certificate authority, email, and other internal systems.

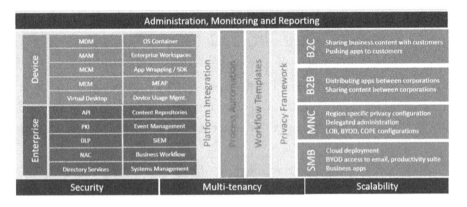

Cloud Connector integrates with the following internal components:

- Email Relay (SMTP)

- Directory Services (LDAP/AD)

- Microsoft Certificate Services (PKI)

- Simple Certificate Enrollment Protocol (SCEP PKI)

- Email Management Exchange (PowerShell)

- BlackBerry Enterprise Server (BES)

- Third-party Certificate Services (On-premise only)

- Lotus Domino Web Service (HTTPS)

- Syslog (Event log data)

Mobile Access Gateway/Tunnel

The Mobile Access Gateway (MAG) provides a secure and effective method for individual applications to access corporate sites and resources. When your employees access internal content from their mobile devices, the MAG acts as a secure relay between the device and enterprise system. The MAG is able to authenticate and encrypt traffic from individual applications on compliant devices to the back-end site/resources they are trying to reach.

Use the MAG to access:

- Internal document repositories and content using the Content Locker

- Internal websites and web applications using the Secure Browser

- Internal resources via app tunneling for iOS 7 and higher devices using the tunnel

On-Premise Configurations

When deployed within an organization's network infrastructure, Workspace ONE can adhere to strict corporate security policies by storing all data on-site. In addition, it has been designed to run on virtual environments, which allows for seamless deployments on a number of different setups.

It can be deployed in a variety of configurations to suit diverse business requirements. Common deployment topologies include single-server, multiserver, and hybrid models. The primary difference between these deployment models are how its components (Admin Console, Device Services, Content Service, Database Server, Secure Email Gateway, Cloud Connector, and MAG) are grouped and how they are positioned within the corporate network.

While three common permutations are further detailed in the following sections, the solution is highly customizable to meet your organization's specific needs. If necessary, contact VMware to discuss the possible server combinations that best suit your organization's needs.

Most typical topologies support reverse proxies. A reverse proxy can be used to route incoming traffic from devices and users on the Internet to the servers in corporate network. Supported reverse proxy technologies include Bluecoat, Microsoft, F5 Networks, IBM, and Cisco. Consult your VMware representative for additional support for technologies not listed here, as support is continuously evolving.

Basic/Single App Server Deployment

This delivery model can be used for organizations managing fewer than 1,000 devices. This configuration allows for simplified installation and maintenance, while allowing future scalability and flexibility as deployments grow. A single server deployment allows for easy integration to enterprise services, as well as simplified control and validation over the entire environment.

Hybrid Server Deployment

A hybrid-server deployment model is recommended for organizations managing between 1,000 and 5,000 devices; however it can be used even for deployments of fewer than 1,000 devices. This configuration differs from the single server model by separating the secure email gateway (SEG) and the database server each onto separate servers. The advantage of this topology comes in segregating the email management infrastructure to be maintained and scaled independently, as well as isolating the database server for ease of troubleshooting and future scale.

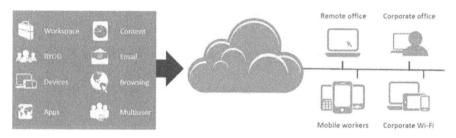

Multiple Server Deployment

A multiserver deployment model is recommended for organizations managing 5,000 or more devices and/or those wanting to utilize a DMZ architecture to segment the administrative console server into the internal network for increased security. This deployment model allows for increased resource capacity by allowing each server to be dedicated to workspace components. While these components are combined, they can also reside on a dedicated server. Many configuration combinations exist and may apply to your particular network setup. Please contact VMware and schedule a consultation to discuss the appropriate server configuration for your on-premise deployment.

Summary

Business mobility is the core of workspace innovation. It takes advantage of advances in mobile technologies and mobile styles of work to create a new business environment where employees can be more effective. It gives organizations the power to be more competitive. Business can work more effectively and connect with customers in ways that once seemed impossible. Business mobility initiatives are typically built upon a digital workspace that allows these business process investments to be more successful. From the constant proliferation of mobile devices to new mobile form factors, few trends will make as big an impact on business operations. VMware understands this and helps customers to reimagine their business processes and change how they compete in the market.

CHAPTER 5

Enterprise Mobility Management

In this chapter, we will talk about an enterprise mobile management solution from VMware as well as newer technologies built specifically for a mobile and collaborative workforce. *Enterprise mobility management* refers to a shift in work habits, with more employees working out of office and using mobile device and cloud services to perform business tasks on the go. Employees are mobile savvy than they've been ever before; therefore, for organizations it makes sense to embrace enterprise mobility. One out of every three employees is completely mobile, and one out of two employees say they are expected to get work done no matter where they are. It is imperative for business and IT leaders to come together and build a strategic and sustainable enterprise mobility strategy that will transform their business to a mobile-centric environment without comprising the customer experience.

Purpose and Intended Audience

This chapter provides detailed information about the end-user requirements to access their critical apps on mobile, tools, and external services required to successfully implement the VMware Workspace ONE solution.

© Ajit Pratap Kundan 2025
A. P. Kundan, *Intelligent Automation with End-User Computing Solutions*,
https://doi.org/10.1007/979-8-8688-1312-2_5

This chapter is intended for EMM administrators and VDI administrators who are familiar with and want to use VMware software to deploy and manage various kind of mobility devices.

Gone are the days of client-server computing when Windows ruled and end users were tasked with doing their work from one device and one location. Today, end users are leveraging new types of devices for work, accessing Windows applications alongside non-Windows-based applications, and they are more mobile than ever.

In this new mobile cloud world, managing and delivering services to end users with traditional PC-centric tools has become increasingly difficult. Data loss and image drift are real security and compliance concerns. And organizations are struggling to contain costs. Workspace ONE provides IT with a new streamlined approach to deliver, protect, and manage Windows desktops and applications while containing costs and ensuring that end users can work anytime, anywhere, on any device. And VMware is uniquely positioned to provide customers with a unified solution that addresses all of their end-user computing needs, from mobile and social to the desktop, to ensure that the virtual workspace becomes a reality.

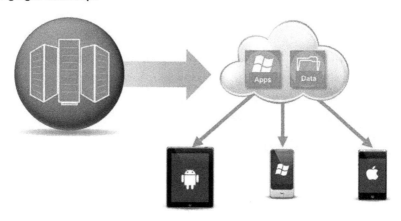

From Client/Server to Mobile Cloud
The changing IT landscape

Solution Scalability Across Different Infrastructure

VMware Workspace ONE enterprise mobility solutions is designed to scale across various infrastructure setups, from on-premises deployments to public/hybrid clouds, ensuring secure and efficient management of mobile devices and endpoints.

- Advanced integration and partner ecosystem

- Common development platform

- Broadest mobility solution set

- Multitenant, highly scalable architecture

- Flexible delivery: cloud and on premise

It provides a secure way for clients to connect to the different resources of an organization. To do so, Workspace ONE needs to be secure. There is a set of built-in security features and various compliances that are adhered to. However, if the customer has any specific requirements in terms of compliance that needs to be adhered to, these requirements can be taken up on a case-to-case basis, by connecting to the support team. Using agent and session host server agent enables the administrator to log user logon events and applications accessed by the user in a remote session for audit purposes. The information collected in a user session is sent to the reporting server to generate appropriate reports/graphs to analyze user activities in remote sessions.

Solution Overview

What BlackBerry Enterprise Server did for BlackBerry devices, VMware does for today's iOS, Android, Windows, and other platforms.

Plan for Extended Enterprise Collaboration

- Suppliers, vendors, and contractors
- Dealers and franchises
- Customers
- Prospects
- Auditors and governance bodies
- Students

Protect Your Corporate Data

We (Solution Engineers/Service Providers) have to protect corporate data, including sensitive and high-value content like customer data, medical records, board books, sales presentations, textbooks, etc. A comprehensive data protection strategy must be in place specially across mobile endpoints, cloud apps, and enterprise systems.

Develop a Comprehensive Mobility Strategy

Manage the Device

Manage the Workspace

Apps   Content   Email   Browsing

Mitigate Business Risks

- Require users to accept the terms of use to access corporate services

- Inform users about data captured and actions allowed on the device

- Track, report on compliance, and update agreements over time

- Assign and enforce different agreements based on the following:

    - User role: End users versus administrators

    - Ownership: Corporate versus employee

    - Platform: iOS versus Android

    - Department, business unit, or country

- Support multilingual agreements across the company

Protect Employee Privacy

- Ensure privacy of personal data

- Set privacy policies that do not collect personal data

- Set custom policies for employee -owned devices

Define Granular Privacy Policies

- GPS location

- User info

- Name

- Phone number

- Email account

- Public apps

- Telecom data

- Calls

- Messages

- Data usage

Automate Compliance to Scale and Build Policies

Organizations must build compliance policies that dynamically assess and act on device posture to ensure secure, compliant, and scalable Enterprise Mobility Management (EMM) solution. These policies allow you to automatically enforce controls when a device is non-compliant based on factors like encryption status, device model, passcode presence, application list, and more.

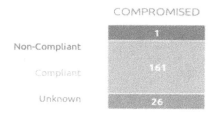

Define Escalation

Time based:

- Minutes

- Hours

- Days

- Tiered actions

- Repeat actions

Specify Actions

- Notify admin when noncompliant

- Send SMS, email, push notification

- Request device check-in

- Remove or block specific profiles

- Install compliance profile

- Remove all profiles

- Remove or block apps

- Enterprise wipe

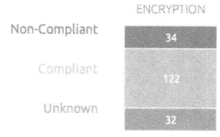

ENCRYPTION

Non-Compliant	34
Compliant	122
Unknown	32

Ensure Regulatory Audit and Compliance

- Log all admin, device, and system events for system monitoring

- Integrate with existing logging tools

- Track the terms of use acceptance

- Determine historical retention requirements

- Adjust policies for local regulations

Severity	Date/Time	User	Module	Category	Event
⚠ Warning	11/18/22 7:43 AM EST	pjohn	App Catalog	Device	App Install Failed
ⓘ Information	11/18/22 7:09 AM EST	tsmith	Admin	Policy	Admin User Edited Passcode Policy
ⓘ Information	11/18/22 6:37 AM EST	sthomas	Admin	User Management	Added to Sales User Group
✖ Critical	11/18/22 6:17 AM EST	mhorton	Device	Device	Device Remotely Wiped
ⓘ Information	11/18/22 6:01 AM EST	srones	Self-service Portal	Login	SSP User Log-in success

139

Enable Self-Service Management

- Reduce IT burden by allowing users basic administration over devices

- Simplify enrollment, configuration, and support

- Enable users to drag and drop files into personal folders in SCL

- Share files with others and set different access and editing privileges

Self-Service User Capabilities

- Enroll additional devices

- Perform remote commands

  - Device query

  - Send message

  - Clear passcode

  - Wipe device

- Download optional profiles

- View device information

  - Compliance audit

  - Installed profiles and apps

  - GPS location

- Request applications and technical support

- Add personal content to SCL

- Enable secure collaboration on content

Manage Strategic Trends with Analytics

- Central portal for fast access to critical information

- View details including:

  Enrollment

  Compliance

  Profiles

  Applications

  Content

  Telecom

  Email

- View device deployment information

- View historical data of company devices

- Provide custom date range views with the ability to drill down to data

- Comparison view

Generate Over 80 Reports | Distribute or Export | Integrate to Business Insights (BI)

Cloud Infrastructure: Trust but Verify

VMware Enterprise Mobility Management (EMM) solution Workspace ONE UEM operates across cloud infrastructure using a "Trust but First Verify" model. This phrase aligns with Zero Trust Architecture (ZTA), where no device or user is inherently trusted and access is continuously validated.

Best-in-Class Technology

Cisco, Dell, EMC, F5, Riverbed, and VMware, ensuring high performance and data security.

Physical Security

Global, world-class data centers feature 24/7, on-site security and operations, badge access, and more.

Data Security

AES 256-bit encryption protects data at rest and in transit.

Enterprise Integration

Integrate enterprise systems with AW Cloud Connector and numerous APIs.

Change Management

Standard evaluation processes and shared decision making eliminate unnecessary changes.

High Availability

Redundant power sources, Internet sources, and data center locations ensure traffic isn't slowed down for your business.

Mobility is Moving Fast, Don't Forget Support

Support Operations: Global offices, 24/7/365 support

Support Team: In-house, trained, multilingual

Technology: Online chat, online knowledgebase, online ticketing system, advanced phone system

Mobile Security Key Features

- Enable multifactor user authentication

- Enforce consistent security policies

- Encrypt sensitive corporate data

- Manage user access to corporate resources

- Support internal PKI and third-party certificates

- Establish network access controls

- Separate corporate and personal data with Workspace to keep corporate information secure

- Enforce compliance rules and escalating actions

- Separate corporate and personal data with containerization

- Remotely wipe enterprise data from device and apps

Mobile Device Management Key Features

- Extend IT security policies to mobile deployments
- Enable access to enterprise services and resources
- Configure device settings and policies through profiles
- Assign profiles based on device, ownership or group
- Quarantine devices and manage by exception
- Automate IT processes and workflows
- Provide helpdesk and self-service to corporate users
- View and report all mobile assets and policies
- Create terms of use based on device ownership type
- Manage rugged devices and printers/peripherals

Mobile Application Management Key Features

- Manage enterprise, public, and purchased apps
- Integrate with App Store, Google Play, and Amazon
- Create a custom enterprise app catalog
- Deploy apps based on user, device role, or smart group
- Enforce compliance with app blacklists/whitelists
- Build advanced enterprise apps using an SDK

- Add security to existing applications with app wrapping

- Track app inventory, versions, and compliance

- Enable single sign-on for enterprise applications

- Run reputation scanning for internal and public apps

Advanced Mobile Application Management

Software Development Kit:

Development and code change required

Enhance applications:

- Register application

- User authentication

- Compromised status

- DLP policies

- Certificates

- Branding

- App configurations

- App tunnel

  - Integration with single sign-on

  - Customize use of SDK for each app

  - Ideal for enterprises with a mobile app team

App wrapping: Development and code change not required

Secure applications:

- User authentication

- Compromised status

- App restrictions

- AppTunnel

- Encryption

- Containerization:

  - DAR encryption

  - Integration with single sign-on

  - Use app wrapping consistently for all apps

  - Ideal for organizations with fully developed apps

App Wrapping Available for iOS and Android apps

1. Start with a developed internal app.

2. Wrap your app from the Workspace Console within minutes.

3. Workspace security features are added to your app and configurations can be updated over the air.

4. Deploy your application to mobile devices via app catalog or push required apps.

Software Development Kit Available for iOS and Android apps

- **Register application**: Authenticate applications with the Workspace agent

- **User authentication**: Enable users to log in using LDAP or Workspace credentials

- **Compromised status**: Detect compromised devices and prevent access to apps

- **Data loss prevention (DLP) policies**: Restrict print, copy/paste and opening of corporate data

- **Certificates**: Secure access through certificate-based authentication

- **Branding**: Custom brand the app according to corporate standards

- **App configurations**: Update settings/policies over the air without code changes

- **App tunnel**: Reach resources behind a firewall using the Mobile Access Gateway

Integrate Workspace Features into Custom-Built Apps

1. Workspace sends the SDK to the corporation.

2. IT develops and codes the SDK into their applications.

3. Create and assign SDK profiles to the integrated
    apps for OTA configuration.

4. Assign the application to devices that are enrolled in
    Workspace.

5. Deploy your application that is secured and
    integrated with Workspace.

App Reputation Scanning

- Scan apps directly in the Admin Console

- Categorize app risks by area

- Privacy

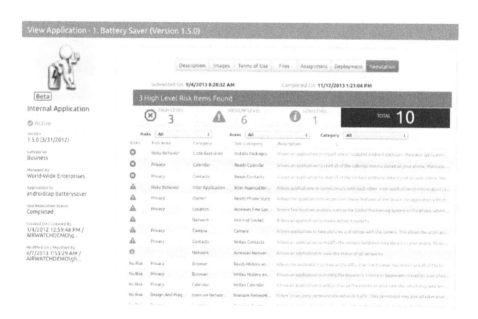

Reputation Integration Partners

- Risky behavior
- Design and programming
- View results by risk levels
- High, medium, and low
- Take action to blacklist unsafe apps

Mobile Content Management Key Features

- Share corporate documents via Secure Content Locker
- Enable two-way sync from desktop to device via Sync service
- Create a corporate container for documents
- Integrate with cloud or on-premise content repositories
- Enforce multifactor user authentication
- Manage user access rights and file privileges
- Delegate admin control across internal groups
- Track document versions, updates, and usage
- Integrate user generated content from the self-service portal
- Allow collaboration through personal folders in SCL
- Enable app-to-app collaboration with single sign-on
- Enable editing and annotation of documents

Mobile Email Management Key Features

- Integrate with enterprise email infrastructures

- Enable a native user experience with Inbox

- Automate configuration of settings and credentials

- Define email compliance policies and actions

- Block email access based on make, model, or OS

- Install, remove, and manage email certificates

- Encrypt email attachments for data loss prevention

- Prevent copy/paste of data to third-party apps

- Wipe attachment content from compromised devices

- Monitor corporate email activity 24/7/365

Mobile Browsing Management Key Features

- Enable browsing through AirWatch Browser

- Configure custom whitelists and blacklists

- Connect to corporate intranet without device-level VPN

- Block native browsers and public browser apps

- Authenticate users with corporate credentials

- Create and require custom terms of us acceptance

- Disable the ability to copy or print

- Define cookie acceptance and clear cookies upon exit

- Enable tabbed browsing and pre-defined bookmarks

Laptop Management Key Features

- Support macOS and Windows PC laptops

- Enroll laptops through the agent or URL

- Authenticate users with corporate credentials

- Deploy preconfigured profiles and products

- Enable scalable distribution with imaging capabilities

- Sync, share, and manage personal content

- View a complete inventory of all connected devices

- Send commands to request information and perform actions

Multiuser Management Key Features

- Configure a single managed device to be used by multiple users

- Assign profiles automatically based on user permissions

- Allow for seamless check-in, check-out process in the agent

- Require users to accept a terms of use (TOU) agreement

- Authenticate users during login with corporate credentials

- Enable single sign-on for access to other apps after login

- Manage devices even when device is not checked out

- Auto-configure single-app mode when no user is logged in

- Configure a time interval for automatic logout

Telecom Management Key Features

- Collect and view real-time telecom data in dynamic dashboard

- Track roaming, data usage, calls, messages, and cellular status

- Protect employee privacy with customized data collection settings

- Monitor plan information and usage trends

- Set automatic actions to remedy excessive telecom voice, message, or data usage

Bring Your Own Device (BYOD) Management

Enable Device Choice

- Support all operating systems and latest device models

- Enable employees to choose the most productive device

Enroll Devices Easily

- Enroll personal devices into AirWatch

- Select "Employee Owned" device ownership

Manage Personal Devices

- Configure policies and settings based on device ownership

- Secure access to enterprise resources, apps, and documents

154

Protect Employee Privacy

- Configure what info is collected based on device ownership

- Isolate and protect both corporate and personal information

Mitigate Business Risks

- Enforce custom TOU agreements for employee devices

- Specify the info being collected and actions IT is allowed to take

Enable Self-Service Management

- Simplify enrollment, configuration, and support capabilities

- Locate, lock, and perform an enterprise or full device wipe

Workspace

- Containerize enterprise data within AirWatch Workspace

- Secure connectivity to enterprise systems

- Consistent user experience with branding, single sign-on, and customizable settings

- Managed only at application layer (MAM); MDM optional

Enterprise Integration

AirWatch Mobile Access Gateway (MAG)

- Enable mobile device access to enterprise systems, corporate email, content repositories, and corporate networks

- AppTunnel VPN between work applications and enterprise systems

- Limit enterprise systems access to corporate apps and configure access by user and device

Directory Services Integration

- Leverage existing directory services and corporate identities

- Import directory structure or select specific groups to import

- Support 1-N user groups through multitenant architecture

- Map existing group assignments with Workspace user groups

- Detect changes in the directory system and sync with Workspace

- Delegate authority to IT admins to manage specific groups

- Assign device profiles, apps and content based on group membership and ensure users receive the appropriate access and restrictions

- Remove access and retire devices as users are deactivated

- Integrate with cloud and on-premise deployments

Certificate Integration

- Enable client authentication and encryption to enterprise systems through certificates

- Integrate with existing PKI or third-party providers

- Secure integration (not dependent on SCEP)

Platforms Supported				
Email	iOS	Android	WP8	
Wi-Fi	iOS	Android		OS X
VPN	iOS	Android		OS X
Web Apps	iOS	Android	WP8	OS X
Apps (SDK)	iOS	Android		
S/MIME	iOS	Android		

- Issue, renew, and revoke certificates over the air

- View all certificates issued across devices

- Renew expiring certificates automatically

- Revoke retired/compromised certs automatically

Email Integration

Proxy Model

- Install Secure Email Gateway (proxy) on premise through a self-configurable setup utility

- Scalable, highly available, and redundant architecture

Microsoft Direct Integration

- Integrate with email infrastructure using Windows PowerShell APIs

- Secure cloud integration via Cloud Connector

Google Direct Integration

Integrate with email infrastructure using Google APIs

Only managed devices receive Google App credentials through automatic provisioning of a profile

Network Access Control (NAC) Integration

- Allow employees to remotely access corporate resources

- Work uninterrupted with automatic connection

- Maintain security with corporate credentials

- Support multiple NAC appliances to connect to a single server

- Include Verbose API for devices, users, profiles, and apps

- Enable event notification (callback) API with no need for polling for device information

- Integrate with the REST API interface

- Recognize and redirect unknown devices to enroll

Content Repository Integration

- Synchronize network shares, file servers, and file systems with Secure Content Locker

- Distribute and synchronize files from enterprise or cloud servers to a device

- Utilize existing corporate credentials for user access

- Provide author, keywords, date created, date modified, versin, notes, created by, etc., in document information

- Use access control lists (ACLs) for user permissions

- Secure distribution from SharePoint without VPN (using MAG and Cloud Connector)

- Apply default settings for security and deployment

System Information and Event Management Integration

- Record and view all device and console events in the console

- Configure device and console logging levels

- Filter events by severity level, category, or module

- **Device events**

  - All communications to and from a device

  - Interactions including MDM commands and responses

  - Information including end-user actions on the devices

Console events

- Console login/session events (including failed login attempts)

- Admin actions for user and device management (including changes to profiles, apps, content)

- Changes to system settings and configurations

- User preference and navigation changes

API Integration

- **REST SOAP integration**: Allow external programs to invoke core product functionality and integrate with existing IT infrastructure and third-party applications

- **Business processes**: Enrollment Authentication Device details Device profiles Applications User/device groups

- **Enterprise systems**: System management Operations management Service management Proprietary systems

Enterprise Architecture

Scalability

Deploy 10 to 100,000+ devices through a configuration that easily scales to support additional device capacity.

Multitenancy

Absorb fragmentation within your corporate infrastructure into a single instance of Workspace ONE.

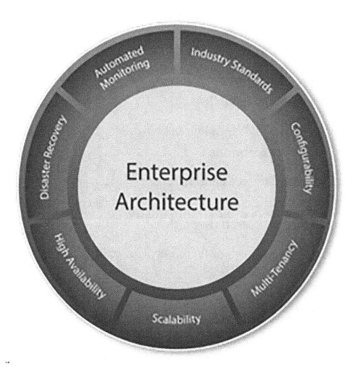

Configurability
Deploy the specific software to your environment and architecture requirements.

High Availability
Deploy in a highly available environment with all components made to instantly fail over without downtime.

Disaster Recovery
Set up software in a remote data center and enable it in the event of a data center failure.

Automated Monitoring
Automate monitoring through a direct plug in to Microsoft's System Center Operations Manager (SCOM).

Scalability

- Take advantage of flexible solution that grows with you

- Support an unlimited number of devices without losing management capabilities

- Apply global settings and organization group settings and allow exceptions

- Manage all devices and processes from a single instance admin view and delegate access to administrators and users

- Utilize either horizontal or vertical scaling

High Availability and Disaster Recovery

- Use active-active configurations for high availability and redundancy

- Utilize load balancers or virtual machines for high availability across multiple data centers

- Protect data through SAN data replication, data backups, and SQL log shipping for a guaranteed quick recovery

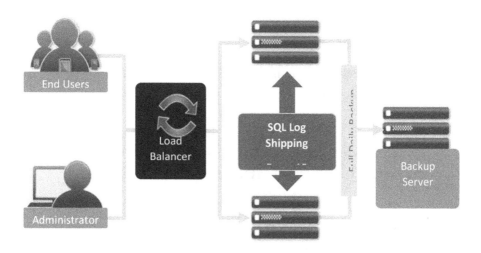

Role-Based Access and Authentication

Choose from 2,000 unique security permissions to define custom roles.

- Set permissions to view, write, or update the system
  with advanced options to define user roles

Permissions	Default Roles		Custom Roles
	Administrator	Help Desk	App & Content Manager
Dashboards	Yes	Yes	
Reports & Analytics	Yes	Yes	
Profiles & Policies	Yes		Yes
Applications	Yes		Yes
Content	Yes		Yes
Users/Admins	Yes		
System Configuration	Yes		
Advanced Features	Yes		

- Authenticate users with basic, LDAP, or SAML credentials

- Auto-assign roles to individual users or groups with LDAP integration

- Allow users to have multiple assigned roles

- Automatically sync roles in Workspace ONE when directory services group changes

Customer Tools and Services

Professional Services

- Purchase a fixed number of professional services and training hours

- Use professional services and training hours as needed

Custom Consulting Services

- Build custom service packages based on your needs

- Assistance with implementation of special business use cases and advanced features

Health Check Packages

- Ensure your Workspace ONE deployment is ideal by performing technical and functional reviews

- System review before go-live or assessment and optimization post go-live

Quick Start Packages

- Get started with Workspace ONE quickly and easily

- Onboarding packages to assist with prerequisites, installation, and initial training

VMware Workspace ONE is uniquely positioned to provide customers with a unified solution that addresses all of their end-user computing needs, from mobile and social to the desktop, to ensure that a virtual workspace becomes a reality.

Summary

Enterprise mobility improves productivity by providing the required information on the go. With more focus on customer responsiveness and employee productivity, it is necessary to take your business to the next generation of technology usage. Mobile devices can also integrate with cloud-hosted services such as Exchange Online, which provides users with the ability to respond to schedule changes and new emails on the fly. When decision-makers are able to communicate approvals to employees, regardless of location, tasks can get done quickly and efficiently.

SDK for Android - VMware Workspace ONE UEM: https://docs.vmware.com/en/VMware-Workspace-ONE-UEM/services/WS1_SDK_Android_Doc.pdf

Recommended Architecture - VMware Workspace ONE UEM 2212: https://docs.vmware.com/en/VMware-Workspace-ONE-UEM/2212/WS1_Recommend_Arch.pdf

High-Level Horizon-Wxorkspace ONE Access Integration Design: https://docs.vmware.com/en/VMware-Workspace-ONE-Access/21.08/ws1-access-resources/GUID-B8B2C3C9-7DBE-4DC4-8F49-FF616DE820E2.html

VMware Workspace ONE Hub Services Documentation: https://docs.vmware.com/en/VMware-Workspace-ONE/services/intelligent-hub_IDM.pdf

CHAPTER 6

Any App from Anywhere on Any Device

In this chapter, we will discuss the desktop-virtualization solution from VMware called Horizon as well as newer technologies built specifically for a mobile and collaborative workforce. We will also discuss how these technologies together enable IT to optimize its current environment while safely embracing innovation and emerging trends to maintain a productive workforce and secure business environment. We will also see how Horizon accelerates application deployment and simplifies application migration with agentless application virtualization and also provides secure access to applications and data on any mobile device or computer, enhancing the end-user experience while reducing management costs. The software can simplify management, security, and control of desktops while delivering the highest-fidelity experience of desktop services to any device, on any network.

It's all about empowering a more mobile, socially aware workforce at the top of the stack, in the Access layer. It's not about devices or machines but what we can do for users themselves. We have seen Horizon continue to grow faster than the VDI industry as a strong testament to the capability, scalability, and manageability it provides to VDI environments. Now with

© Ajit Pratap Kundan 2025
A. P. Kundan, *Intelligent Automation with End-User Computing Solutions*,
https://doi.org/10.1007/979-8-8688-1312-2_6

Mirage added to the overall solution of how to centrally manage, secure, and ensure continuity of the traditional desktop environment, Horizon will become the broker for the cloud era.

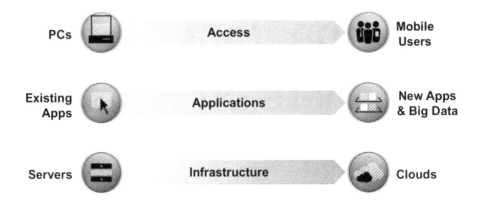

First let's look at how VMware is helping **transform legacy desktops into a managed service**. As workers become increasingly mobile, IT must provide anywhere, anytime access to the traditional Windows desktop.

- VMware Horizon continues to grow faster than the VDI industry for managing virtual and physical desktops, as a strong testament to the capability, scalability, and manageability it provides to VDI environments. Hosting desktops in the data center using VMware Horizon comes with many benefits including simplified systems and application management, increased security and control, and higher availability and agility.

Cisco, Dell, HP, and Intel joined the growing ecosystem of OEM, technology, and channel partners that are helping to develop and sell VMware Horizon appliances. These easy-to-deploy VDI appliances host between 25 and 500 virtual desktops that can be provisioned quickly, making them ideal for small and midsize businesses (SMBs) and scalable for larger organizations.

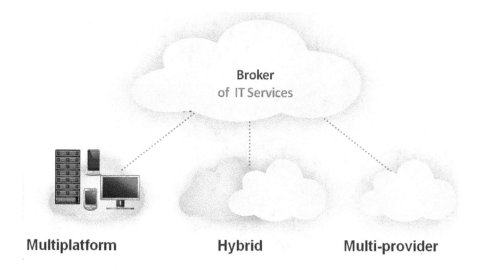

Multiplatform   **Hybrid**   **Multi-provider**

VMware Horizon Solutions

VMware introduced new solutions designed to help IT administrators best architect their environment and ensure all components selected work together for quick managed desktop deployment. VMware Horizon Business Process Desktop and Branch Office Desktop have been tested to ensure product interoperability and performance and are supported with reference architectures, tools, and services from VMware and VMware technology partners to minimize guesswork around VDI.

VMware and its partner community can now offer centralized desktop management solutions that address the requirements of IT organizations to help them transform legacy Windows desktops into a service. VMware allows local system execution on physical desktops and laptops. It clones the images of the endpoints in the data center and runs them locally. This gives customers all the benefits of centralized management and recovery while allowing users to work offline and preserve an uncompromised user experience.

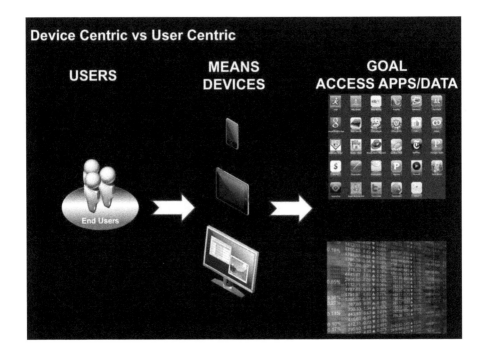

VMware is committed to bringing the solutions in both of these areas into one cohesive platform for the post-PC era.

- We start with simplification by turning the discrete components into services that can be centrally managed, updated, and secured.

- We can get rid of all of these different tools and all these point solutions and provide something that's integrated and extensible to support what we have today for IT, as well as all the new technologies coming from VMware and from other partners along the way.

- Ultimately IT is getting centralized management to enforce policies, which in turn help them be compliant and safe with corporate data.

- We then provide a universal services manager to enable policy-based settings for user access to these services through Horizon.

- Finally, we deliver the appropriate content to any device, anywhere, anytime, based on policy settings. If the user needs a full desktop, they get access to their virtual desktop. If they just need an app, then they just get an app, preserving the user experience of the endpoint device.

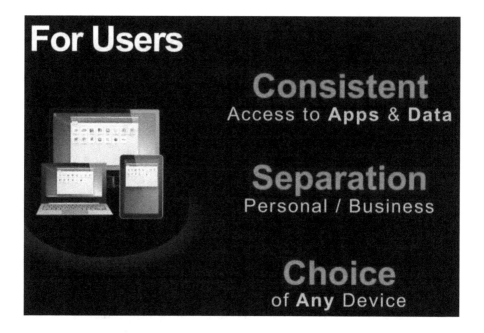

- This is to support the full embracing of the post-PC era.

  The new Horizon Suite provides value to two sides of the equation, which almost never happens in an IT solution.

- As a user, you get consistent access to all your applications and data from whatever device you happen to be on. You have a clear separation of personal and work life so that you can make sure you are compliant and doing the right things, and then you have a choice of lots of different devices.

- VMware wants to give you the best environment for your particular device that you have with all these different delivery mechanisms.

Horizon Suite Is Important to Customers

- **Simplify**: Streamline and simplify operations by turning disparate operating systems, applications, and data into centralized services that can be easily provisioned, managed, and delivered to end users.

- **Manage and secure**: After end-user assets are transformed into centralized services, they can be managed, secured, backed up, and kept current from a single place. Policy-driven access and delivery safeguard vital data and ensure compliance.

- **Empower**: Mobile users need the freedom to choose the right device for each task and setting. Galvanize workforce productivity by delivering a consistent, intuitive, and collaborative computing experience across all devices—anytime, anywhere.

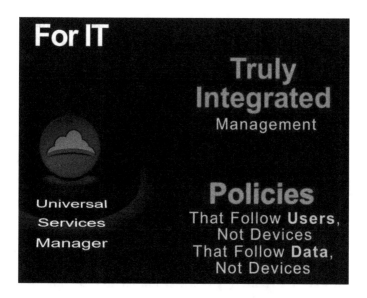

- Horizon Suite allows you to increase the size of deals and profitability. Horizon Suite allows you to partner with customers on key strategic initiatives around workplace mobility, application modernization, and data collaboration.

- Selling Horizon Suite allows you to sell additional services and infrastructure to help customers assess, roll out, and manage their deployment.

Better End-User Experience

- Horizon Suite has support for dedicated graphics acceleration with Nvidia and ATI to ensure developers and power users can leverage high-end 3D graphics in their virtual desktop. Horizon Suite complements VMware's support of shared graphic acceleration and non-hardware-accelerated 3D graphics to ensure that customers have a wide range of performance and price points to choose from.

- Horizon Suite has support for Windows and multimedia redirection in Windows desktops to ensure smooth multimedia streaming with minimal impact on bandwidth.

Streamlined Management

- Support for vSAN to allow customers to take advantage of VMware's software-defined storage tier to pool the compute and direct-attached storage resources, automate storage provisioning, and drive down the costs associated with storage around VDI.

- Support for ThinApp, which includes support for 64-bit architectures.

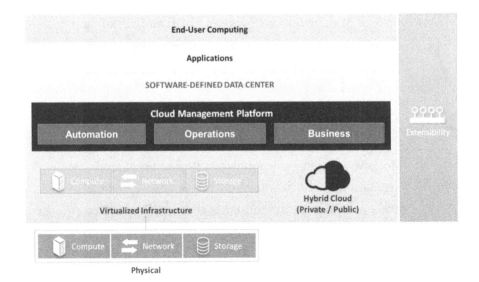

- Agent Direct Connection supports distributed implementations of Horizon with highly resilient desktops that are no longer tied to the WAN. With a connection, users can log into their session without having to authenticate into a connection server and enjoy highly responsive, high-performance desktops even in locations with poor bandwidth connectivity.

- The delivery models is desktop as a service.

- Support for Windows Server reduces the cost, complexity, and licensing restrictions associated with supporting cloud-hosted virtual desktops with multitenant infrastructure. For on-premise deployments, Server OS support gives customers greater choice around supporting non-Windows-based devices and in deciding which Windows licensing model makes the most sense for their use cases.

Use Cases

Legal firms deliver desktops to personal devices while reducing overall IT costs and improving productivity.

Problem Statement

- Desire to enable BYOD to reduce hardware costs and stop managing hardware

- Desire to find a way to maintain high levels of IT service in light of IT staff reductions

Solution

- Deployed the Mobile Secure Workplace solution with VDI

- Chose VMware due to its strong virtualization relationship and for VMware's depth of knowledge and clear, articulate vision for end-user computing

Benefits

- Reduced number of "tech touches"

- Reduced provisioning times from four hours to ten minutes

- Improved uptime and availability for billable attorneys

- Cut mobile device costs

VMware EUC moved to desktop-as-a-service, which is easier for IT to manage. Lawyers have more flexible access to the desktops. No matter where they are, they get the exact desktop they have in the office.

Virtual desktops at financial operations organizations go the distance to deliver high performance to streamline operations.

Problem Statement

- Offshore financial operations to remote places without building any local infrastructure or data

- Deliver a high-performance desktop experience to employees in a different country

Solution

- Deployed the Business Process Desktop solution with Horizon on zero clients

- Chose Horizon due to the existing success with VMware for data-center virtualization

Benefits

- Lower financial OPEX costs and project is the model for worldwide operations

- End users cannot tell the desktops are remote

- Eliminate data security challenges

- Eliminate spyware and blue screens of death

We've been able to build a system that is completely dynamic, manageable and scalable and deliver it thousands of miles away. No one can tell that it's not running locally.

Take flight with virtual desktops to save IT time and money while improving the end-user experience and increasing security.

Problem Statement

- Poor control of a fleet of four generations of PCs that crashed regularly with no backups resulting in massive data loss

- Overworked help desk during business hours, which was driving up their costs

Solution

- Deployed the Branch Office Desktop solution with Horizon to hundreds of desktops across all offices

- Solution met all of their requirements and supported their mission-critical reservation application

Benefits

- Achieved 90% savings on power alone by switching to zero clients.

- Een with the addition of the server power in the datacenter, saved a cumulative 60% to 70% (90% solely on the endpoint side).

- 18-month ROI solely on power savings by switching to thin clients (travel costs of IT techs have not yet been included in ROI calculation).

- 100% virtualized on the server side and reservation system = mission-critical application does *not* work in Terminal Services/Presentation server environment. Inside of a VDI desktop, reservation system app runs more quickly and performs better than before when installed on physical desktops.

- 50% reduction in help-desk costs.

- Lockdown endpoints and secure data in the data center.

- Mission-critical app, like reservation system app, performs better in virtual desktop.

It seems utterly absurd when you first hear that only three employees are supposed to manage hundreds of live desktops workplaces distributed across the country. But it is true! Our experience with Horizon has made me completely confident in the solution we have selected.

Financial credit organization delivers a familiar desktop at lower costs with better security.

Problem Statement

- Existing app presentation solution on thin clients was difficult to configure, support, and experienced daily failures

- Needed a more stable end-user environment to manage costs

VMware Solution

- Deployed the Branch Office Desktop solution with Horizon across branch offices

- Horizon cost less, used less RAM, and delivered a more stable desktop

Benefits

- Easy rollout due to familiar desktop experience

- Ability to deliver training videos across sites

- Quickly fix security issues and viruses

- Reduce IT time spent on provisioning and updates

VMware Horizon solution is like a breath of fresh air. It works like it's supposed to. We have time to focus on enhancing the network and supporting the business strategically instead of holding it back. Citrix XenApp replacement (unstable, more expensive, difficult to configure and manage). Branch employees at different sites, Saved money in RAM compared to Citrix solution.

Journey to Horizon Begins Here

These are just a few examples of the tremendous benefits that come with using VMware Horizon. Since it is built to help you improve what you have now, there's no reason you can't start making the transition today.

Here are some steps to get you started:

1. Identify your potential view use cases.

2. Assess the user needs and the environment.

3. Virtualize your applications.

4. Establish "rules" for new apps.

5. Rationalize all desktop images.

6. Begin a VDI pilot and roll to production.

Starts with Apps Orientation

The app landscape is diverse: there are traditional client-server apps and modern cloud-native apps. IT has to manage them all.

- Architecture of underlying infrastructure: SDDC, best architecture for instant, secure, and fluid model of IT

- Abstraction, pooling applied to storage, networking

- One unified hybrid cloud with a common cloud management platform

We call this "one cloud for any application."

Runs on a wide variety of hardware infrastructure without any vendor lock-in situation.

Business mobility solutions enable the safe consumption of apps and content on any device.

Applications are the engine that drive competitive advantage and enable new business opportunity for companies that rely on IT as a competitive weapon.

The demand for new and updated applications means shorter release cycles and the adoption of new techniques like agile development process designed to help app dev groups keep up with increasing demand.

Slow Infrastructure Service Delivery Times

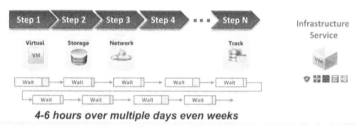

4-6 hours over multiple days even weeks

- Organizational siloes lead to wait times in slow workflows
- Time consuming manual tasks
- Manual configurations lead to inconsistencies, errors and need for rework

... but even with infrastructure automation what about the application?

This pressure is also felt by the IT Operations group who are faced with increased demand and faster delivery expectations for compute resources and services to support both app dev and production environments. The result is dissatisfaction up and down the chain and a growing need to look for alternatives that can deliver timely solutions and reduce risk. Let's talk about innovations and new architecture for IT.

Policy-Based Governance: Personalize the Cloud

- Personalize services using policies to meet unique business and IT needs

- Access and delivery policies to provide the right-size service at the right service level

- Rapidly configure policies with a few clicks

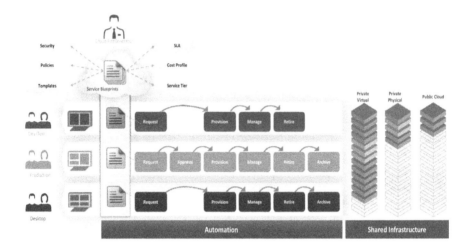

Serving apps efficiently all starts with the architecture of the underlying infrastructure and data center. VMware introduced the concept of a *software-defined data center*. We believe a software-defined data center is the best architecture for this instant, secure, and fluid model of IT.

VMware's architecture for the SDDC builds upon the core concepts that made compute virtualization so radically impactful, namely, abstraction and pooling, and applied those virtualization fundamentals to the other key assets within the data center: storage and networking.

This architecture can be leveraged not just in on-premises data centers or private clouds but also by public clouds. For example, a data center can extend to the VMware public cloud or one of the many managed clouds within the cloud network of service providers. The result is one unified hybrid cloud. All of these environments are managed from a common cloud management platform.

Built upon the software-defined data center architecture, we call this unified hybrid cloud "one cloud for any application." It runs on a wide variety of hardware infrastructure, whether you want to leverage existing infrastructure or adopt converged or even hyper-converged infrastructure. One Cloud offers complete choice.

Now we have talked about how applications are ultimately what users care about. Our users are increasingly expecting to consume their apps on the devices of their choice. Our business mobility solutions enable the safe consumption of applications and content on any device, all effectively managed by IT.

Let us break down the challenge of slow service delivery time by exploring first the challenges for infrastructure service delivery.

Creating infrastructure services is typically a time-consuming manual task. Based on customers surveys conducted, we know that the actual work effort is typically around four to six hours. However, those four to six hours are spread over days or weeks since this effort involves multiple teams, which often operate in siloes. As a result, there are wait times in slow workflows. Moreover, manual tasks lead to inconsistencies and errors in configurations. There is the need for time-consuming rework to ensure that systems consistently; i.e., systems need to behave in the same manner. So now, after a few days or weeks, we provided an infrastructure service, but what about the applications? This is ultimately what the business cares most about. So what is needed to address those fundamental challenges?

In the end, service delivery times need to be significantly accelerated. In other words, a consumers should have the ability to request a service, which then gets delivered automatically within minutes.

Automated IT to IaaS Illustrated

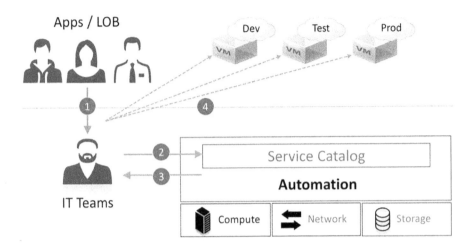

Accelerated delivery of infrastructure services

A service consumer could be the virtualization admin, someone in the IT organization, or perhaps someone in the applications team. However, ultimately it will be necessary to provide the business with the ability to request and provision applications. Applications run on top of the infrastructure services, so the business needs to have the ability to provision the entire stack consisting of application services layered on top of infrastructure services. This is possible only if service delivery can be automated within a single solution, where application services are layered on top of abstracted infrastructure services. A single solution to automate service delivery is the key requirement to reduce service delivery times to minutes.

Accelerating service delivery times with other associated requirements:

- It needs to be possible to rapidly implement the service automation solution. Traditional systems designed for physical environments often required many months or years of implementation, so IT already lost the battle right there. It needs to be possible to integrate into existing environments and flexibly extend them to meet specific business requirements

- Once consumers have the ability to request their own services, e.g. through a self-service catalog, IT needs to ensure that they can still keep control over service provisioning for example who has entitlements to provision services, the service sizing, etc.

- IT also needs to ensure efficiency for delivered services. IT needs to ensure that services are not over-provisioned, or unused services can be identified and reclaimed to contain costs.

- Services need to be standardized. This is important because otherwise automation can never be effectively accomplished. Standardization not only includes eliminating manual service configuration efforts and moving to standardized configurations but also standardizing the software stack, standardizing software versions, etc.

- IT organizations can no longer focus solely on the delivery of IT services. As they move to cloud, IT organizations need to transform their role into "brokers of IT services," operating as a strategic partner to the business offering IT as a service. And the cloud is the key technology that will enable this new operating model.

- Many IT organizations are building a private cloud in order to be more agile and responsive. However, IT will need to increasingly rely more on external service providers as well to deliver IT services that scale to meet business demand and respond to market changes. As a result, successful enterprises will be those that understand how to multisource (on-premises/off-premises, private/public/ hybrid cloud) and manage a hybrid/heterogeneous IT environment, integrating it into their business.

- As a broker, IT will enable LOBs to make fact-based decisions on service consumption and provisioning based on real-time visibility into criteria such as the cost, health, security, and compliance status of services.

VMware helps to deliver the foundation for IT-as-a-service with the SDDC, enabling the transition of IT to becoming a true broker of IT services.

The SDDC is the ideal architecture for:

- Building, delivering, and managing business applications

- Building and operating private, public, and hybrid clouds, where all infrastructure is virtualized and delivered as a service, and the control of this data center is entirely automated by software and enables the delivery of IT-as-a-service

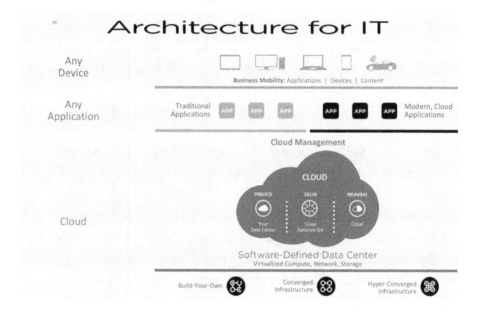

The SDDC enables you to utilize the investments in people, process, and technology that you've already made to deliver both legacy and new applications while meeting vital IT responsibilities. It allows businesses to use what they have today to build for change in the future.

VMware's SDDC solutions provide a unified platform across the hybrid cloud, built on VMware's best-in-class compute, storage, and networking virtualization technologies. It includes the industry-leading cloud-management platform, as well as programmatic management capabilities via OpenStack and infrastructure-level APIs.

The Cloud Management Platform (CMP) becomes the key control layer in an SDDC to manage heterogeneous and hybrid environments. With policy-based automation, operations, and business management capabilities, it helps IT to deliver on the new business need for speed, agility, and choice while also delivering the ongoing IT need for control and efficiency.

Traditional tools and processes that infrastructure and operations teams are using aren't optimized to deliver on the promise of the software-defined data center and hybrid cloud. Businesses need a cloud management platform that simplifies and accelerates infrastructure and application delivery and ongoing management onsite in their data center and in the cloud.

VMware's CMP is purpose-built for the hybrid cloud. It provides a comprehensive management stack that enables IT to deliver infrastructure and applications quickly on vSphere and other hypervisors, physical infrastructure, and private and public clouds, all with the control IT needs.

VMware's CMP is distinctive, by offering not only automation but also integrating operations and business capabilities into a single platform to provide performance and financial insight to improve IT/business decision-making and alignment. This helps IT to accomplish its mission as a broker of IT services, by enabling IT to source, provision, and manage the lifecycle of IT services across the new data center landscape: multiplatform, hybrid cloud, and multiple providers.

- **Automation:** Self-service provision infrastructure and applications across multiple hypervisors and private and public clouds with both speed and control.

- **Operations:** Manage infrastructure and applications in physical, virtual, and cloud environments with integrated capacity, performance, log, and configuration management.

- **Business:** Align IT spending with business priorities by getting full transparency of infrastructure and application service cost and quality.

- **Unified Management:** Use a common set of tools across on-premises and public clouds with a unified management experience and fast time to value.

CMP supports cloud administration, server administration, fabric, and operations teams working with lines of business to ensure agility, application SLAs, and infrastructure efficiency with control. The platform combines automated delivery, intelligent operations, and business insight to deliver a unified cloud management experience across hybrid clouds and heterogeneous environments.

VMware Strategy

- Software-defined data center, the hybrid cloud, and end-user computing form the basis of the three strategies for VMware.

- These three together are the path to the software-defined enterprise.

- Execute the software-defined enterprise, building on these three components.

The Software-Defined Data Center: An Open, Industry Architecture

The architecture that provides the competitive advantage sought by today's businesses is the software-defined data center (SDDC). The SDDC is an open architecture that extends the principles of virtualization such as abstraction, pooling, and automation beyond compute to the rest of the data center, bringing deployment and resource management to any infrastructure and cloud environments of choice. This in turn enables the management and orchestration for all IT services needed in the mobile-cloud era.

As the leader in compute virtualization through its industry standard vSphere product, VMware is the company best positioned to extend virtualization solutions to the rest of the data center. VMware's SDDC solutions deliver cloud service provider economics in the data center, as well as fast and agile application provisioning that responds to ever-changing business demands. VMware's SDDC solutions offer the right availability and security for each application with policy-based governance, as well as the ability to run new and existing applications across multiple platforms or clouds.

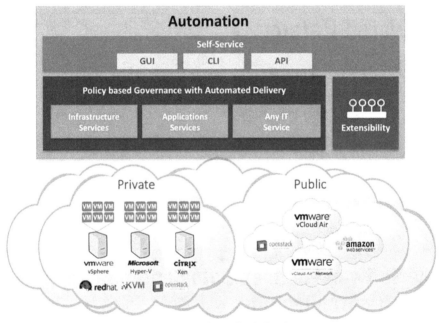

Multi-Vendor Hybrid Cloud

Multivendor hybrid cloud infrastructure: choice, flexibility, and control

VMware's Software-Defined Data Center Solutions consists of four components:

- A completely virtualized infrastructure across compute, network, storage and automation

- Delivered on- or off-premises in a hybrid cloud;

- A comprehensive virtualization and cloud management platform

- Choice of cloud frameworks, whether a fully integrated VMware stack or OpenStack-based

- Create service offerings that meet the needs of the Dev/Test Group; you can also do the same for Production and Desktop.

- The policies defined in the cloud blueprints not only determine the process steps that will be used to provision and manage each resource but also dictate what type of machine (virtual, physical or cloud) as well as the service level each user will receive.

- All of these policies can be configured by the administrator without having to write any scripts or code.

- Support a **unified service catalog** and App Store ordering experience where users can request infrastructure services from a personalized collection of application. In addition, administrators can use the Advanced Service Designer to automate and make available custom IT services through the new IT service catalog. Service entitlements and optional approvals allow IT administrators to deliver a personalized IT service catalog that can be optimized to the specific needs of individual users or groups of users. Each Tenant can have their own specific branding and user authentication directory services (LDAP).

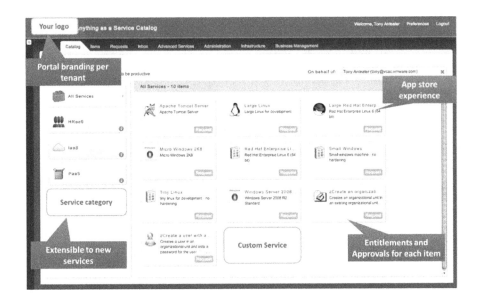

Unified Service Catalog: applications, infrastructure, XaaS, desktops

- Once a consumer requests a service, the service cost is displayed. More service costing capabilities are available on the Business Management tab.

Service Categories

Service categories organize catalog items into related offerings to make it easier for users to browse for the catalog items they need. For example, catalog offerings can be organized into Infrastructure Services, Application Services, and Desktop Services.

Entitlements determine which users or groups can request catalog items or perform specific actions.

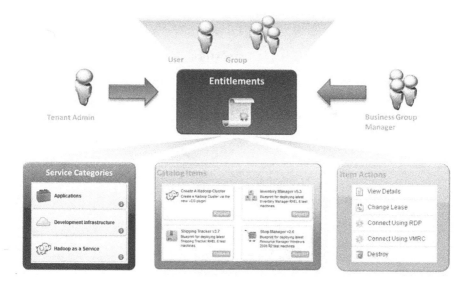

A tenant administrator or service architect can specify information about the service category such as the service hours, support team, and change window. Although the catalog does not enforce service-level agreements on services, this information is available to business users browsing the service catalog.

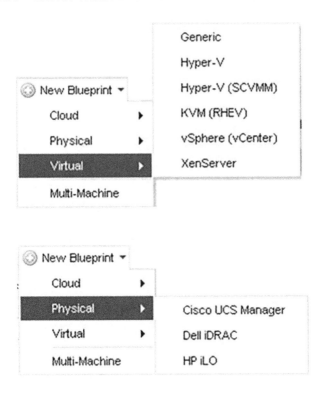

Types of Blueprints

There are four types of blueprints. Each type can deploy machines on different platforms. The four types are as follows:

- Cloud

- Physical

- Virtual

- Multimachine

Catalog Items

Users can browse the service catalog for catalog items that they are entitled to request. Some catalog items result in an item being provisioned that the user can manage through its life cycle. For example, an application developer can request storage as a service and then later add capacity, request backups, and restore previous backups.

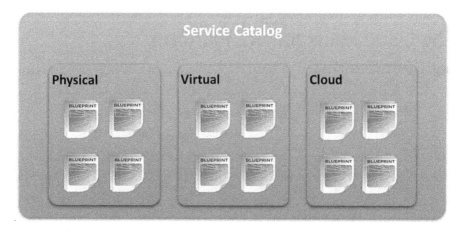

Other catalog items do not result in provisioned items. For example, a cell phone user can submit a request for additional minutes on a mobile plan. The request initiates a workflow that adds minutes to the plan. The user can track the request as it progresses but cannot manage the minutes after they are added. Some catalog items are available only.

Extensibility: VMware can help customers to adapt and extend Automation by calling/integrating external tools and applications during the delivery process for seamless operations.

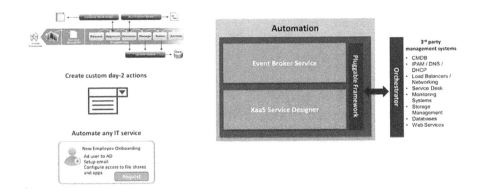

The Cloud Management Platform Customer Journey

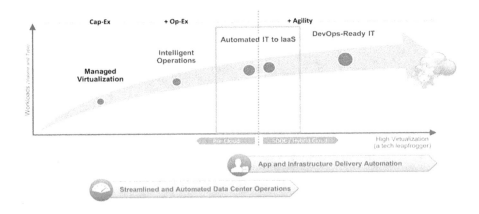

- IT requests services in the services catalog on behalf of consumers.

- IT delivers infrastructure services to apps teams.

- Later, app teams can be entitled to request services on their own.

Summary

VMware Horizon Workspace is an end-user computer solution for organizations that want to simplify IT control while empowering the mobile workforce. Horizon Workspace simplifies the end-user experience and reduces IT costs by combining applications and data into a single enterprise-class aggregated workspace, securely delivered on any device. End users gain freedom of mobility through anytime, anywhere access. For the administrator, the results are simpler, centralized, policy-based management and control of IT consumerization.

CHAPTER 7

VMware's Active Application Management Strategy for IaaS

In this chapter, we will discuss the separation of the application and the infrastructure (cloud provider/cloud consumer model). By moving to a horizontal approach, customers will able to gain a lot more control over the application and adapt it to the business needs (scale and remediation). We will explain the motivation for the separation and focus on the Application layer as a complete operational stack that helps the application owners and developers achieve better agility, more flexibility, more control, higher performance, and lower maintenance costs.

The move toward the cloud and the use of modern application frameworks and methodologies is creating an opportunity to think differently about managing applications. You want to:

- Manage applications holistically and separately from the supporting infrastructure

© Ajit Pratap Kundan 2025
A. P. Kundan, *Intelligent Automation with End-User Computing Solutions*,
https://doi.org/10.1007/979-8-8688-1312-2_7

- Help manage applications that move across, and run across, public and private clouds

- Horizontally integrate management across deployment, monitoring, and change

VMware's Application Management Guiding Principles

Simplicity

Guiding principles of VMware's Application Management within the Workspace ONE platform is designed to ensure secure, seamless, and efficient application delivery across all endpoints (mobile, desktop, rugged, or BYOD).These principles guide how applications are deployed, secured, updated, and retired in modern digital workplaces.

- Simplified deployment

- Quick time to value (minimum configuration overhead, works in minutes)

- Intuitive UI and navigation

- Applied automation

- Heterogeneity and cloud ready

- Support hybrid cloud deployments (private, public)

- Seamless mobility across clouds

- Single console to manage applications running on any cloud

- Active application management

- Early instrumentation during development/deployment

- Automatic update for configuration and code change

- Native remediation and optimization actions

- Enables collaboration between dev,
  Build & Release (B&R), and Operations teams

My Workspace delivers what users want. What do users really want in 2024 ? They want everything.

- Ability to use any app, on any device, any time, from anywhere

- Strong security, but without inconvenience

- Personalized services that recognize each user's uniqueness

- All the support they need, when and where they need it

- Access to all their data, all the time

What our users want is aligned with the strategic direction of customers.

End users and executives benefit from:

- **Higher productivity and efficiency**: The ability to get more work done in less time with less frustration

- **A better experience**: Faster, more flexible access to apps, data, and support services

- **Privacy and protection**: The ability to work with total security and privacy without onerous restrictions or inconvenience

- **Improved decision support**: Better IT outcomes and business processes through better use of data and analysis

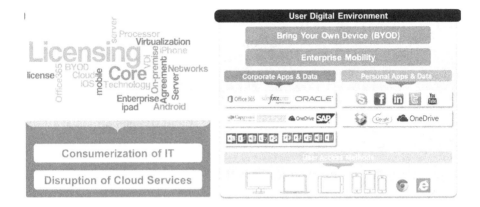

The end-user landscape is increasingly complex

Investment in modern user-centric services pays off in a better user experience with lower costs

STAFF

- Attract and onboard employees

- Employee mobility

- Contractor management

BUSINESS

- Re-org (Reorganization), M&A (Mergers & Acquisitions), rightshore

- Business agility

- Real estate consolidation

- Eco/green efficiency

OPERATIONS

- Branch optimization

- Increase sales productivity

- Improve services delivery

Security

DELIVERY

- Flex-desking and DR

- Application delivery

- Social, mobility, analytics, and cloud

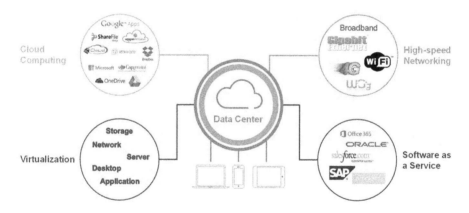

The time is right now because of the following

What Does My Workspace Do?

My Workspace secures access to applications and data on:

- Any device

- Anywhere and anytime

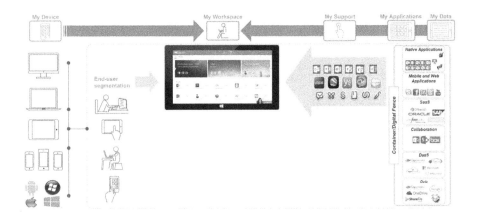

- From any cloud service

- Single admin portal

- Single portal to consume services and get help

- Hosted on-premise with the client or supplier or in a cloud data center

How do we simplify app management?

- Secure, extensible file access platform

- Governed and tracked by IT

- Operable by IT or SP

- For internal and external users

- Accessible anywhere, any device, anytime

- No need for new user behavior

- Collaboration features

- Tightly integrated with existing infrastructure

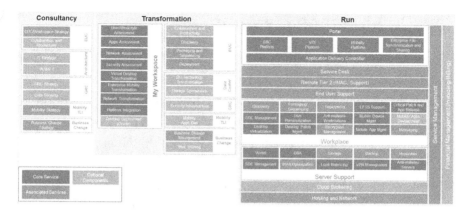

End-to-end service delivery

Cloud Application Performance Monitoring

By moving to an horizontal approach, customers will able to gain a lot more control over the application and adapt it to the business needs (scale and remediation).

The motivation of the separation and focus on the Application layer as a complete operational stack allows the application owners and developers to achieve better agility, more flexibility, more control, higher performance, and ultimately lower maintenance costs.

- IaaS will drive the adoption of the cloud operating model, which defines a clear separation of ownerships.

- It requires a new approach, letting go of some operational aspects in favor of others.

- An horizontal approach is best suited for the cloud operating model.

How Do I Manage My Apps?

A constant theme I hear when talking to customers about managing their virtual, physical, and cloud-hosted environments tends to center around the challenges they face with applications. In particular, the questions that customers are looking to solve are ones like how can I simplify management of my hundreds of apps throughout their lifecycle, from provisioning to updating to dealing with app conflicts? How can I optimize delivery of applications to all the different environments, devices, and users I manage? How can I drive down costs of managing application conflicts or instill a way for easy and simple login for my users? How do I monitor my applications before performance degradation occurs?

Lifecycle Management

>**Provisioning:** Updates, retirement

>**Delivery:** Physical, virtual, and cloud

>**Others Conflict:** Simple login, monitoring

Why is app management so difficult? With VMware, we're trying to drive application management transformation to make your life easier. You may be experiencing many of the issues that come with managing apps such as legacy provisioning, complex and lengthy updates, having to manage separate solutions for app delivery, etc.

Drive application management transformation before:

- Legacy provisioning, complex and lengthy update process, separate solutions for app delivery

- Application conflicts, legacy experience

Drive application management transformation after:

- One-to-many provisioning, deploy apps in seconds, one solution for app delivery, application isolation, SSO app access, consume on any device, performance monitoring

VMware is taking a different approach to managing applications. VMware has methods to provision one app to many desktops, saving considerable time and storage costs. You can deliver apps in seconds and at scale. You can isolate apps to remove the app conflict barrier. Apps can then be consumed on any device and on one portal. You can even transform the way you monitor your applications for better performance.

A New Approach to Application Management

This new approach to app management starts with producing appstacks and layers with natively installed apps, or ThinApps where required. IT can deliver appstacks to virtual and RDSH/published app environments and layers to physical/BYO environments. We support many types of environments, including virtual desktops with VMware Horizon View or Citrix XenDesktop, RDSH, or Published App servers like Citrix XenApp, and physical/BYO environments too. This translates into the end user being able to consume apps and desktops from any device, be it a tablet, a desktop, or a mobile device, and from any location. IT can continuously update layers and appstacks to ensure the latest patches and updates are delivered. And VMware offers a monitoring solution with proactive alerting to help manage your app environment.

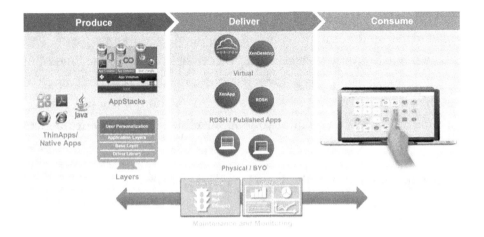

This new approach to app management has helped customers already:

- "Lowers your TCO of packing applications by more than 50%"

- "Decreases storage capacity by 30% when delivering and managing applications"

- "Cuts device management time by nearly two-thirds"

- "Reduces help-desk calls by 75%"

- "Minimizes downtimes for updates and outages for higher business productivity"

- "Optimizes infrastructure to lower CapEx cost"

The following are the technologies available in the app management architecture we're helping our customers with. We provide app isolation, provisioning, delivery, access, maintenance, and monitoring...all across virtual, physical, and RDSH/published app environments. Let's start with focusing on packaging and isolating apps.

Application Packaging/Isolation

ThinApp is all about application isolation and virtualization. Applications and data are traditionally tightly coupled with the OS. This can sometimes cause apps to conflict or cause app compatibility issues, which can cause loss of user productivity. Enter ThinApp. ThinApp uses an agentless approach to virtualization applications so they are decoupled and isolated from data and the OS. A ThinApp packaged app can reside within a Mirage app layer and be delivered via a App Volumes AppStack.

Problem: Tightly coupled relationships between OS, applications, and data

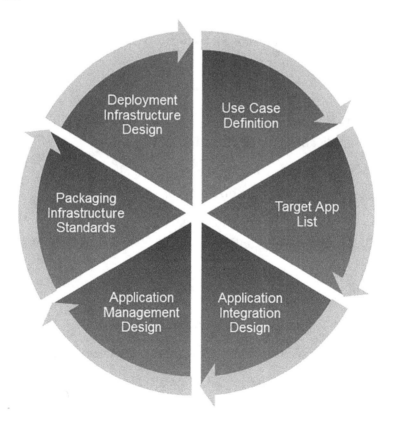

Symptoms:

- Application conflicts

- Complex compatibility test matrices

- Diminished time to deployment

- Loss of user productivity

- Reduced help-desk support calls

Solution: VMware ThinApp

- Agentless application virtualization decouples applications and data from the OS

- Lower your total cost of ownership (TCO) of packaging applications by 50 to 74%

Solution

- Agentless application virtualization decouples applications and data from the OS

- Create conflict-free applications and allow legacy applications to run on newer operating systems

- Optionally provision to virtual and physical desktops, using appstacks and layers

Benefits

- Reduce TCO by reducing or eliminating need to install applications and maintaining central control

- Ensure security without compromising user flexibility

Next up is provisioning and allowing all types of apps to end users.

Application Management Architecture

Application Management Architecture

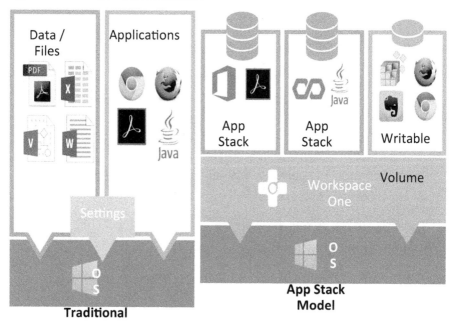

Application Provisioning

Our solution allows customers to deliver RDS hosted apps, session-based desktops, packaged ThinApps, and virtual desktops from a single, secure platform.

RDSH apps and desktops are delivered via PCoIP to ensure a great user experience and are ideal for mobile users who just need access to a few Windows apps across non-Windows-based mobile devices, remote end users who also just want access to a few applications versus a full desktop, BYOD and contract workers, and IT administrators who are looking to streamline day-to-day management by delivering and managing just applications versus full desktops.

Solution

- Provision RDS hosted apps and desktops, virtual desktops, and ThinApps using a single platform

- Full RDS hosted apps support: Windows Media Multimedia Redirection (MMR), scanner redirection, HTML access

Benefits

- Streamline management, with best-in-class security

- Better user experience with optimized performance over the WAN

Let's now focus on real-time app delivery and lifecycle management to virtual and physical environments.

- Reliability, ease of management, security, cost savings, performance…it's not too big a stretch to say that most IT departments equate these values with VMware when talking about Virtual Cloud Infrastructure (VDI).

- But to be honest, VDI (running virtual desktops in the data center) is a lot newer to everyone, so it's not always clear how to get these benefits from any vendor's Virtual Desktop Deployment.

- But that's not to say it isn't on people's minds.

VMware recently commissioned a survey of potential VDI customers across industries and geographies and It's no surprise that the very things that vmware guys come to expect from server virtualization are the top purchase criteria for those considering a VDI purchase. If (VDI) vendors don't live up to these expectations, the result is a lot of frustration and missed opportunities for one's business. So the question is how do get this value out of a Virtual Desktop solution…!

One key piece of the puzzle is the power and maturity of Virtual Cloud Infrastructure itself. VMware can make VDI deployments more performant, cost cutting, secure, reliable, and easy to deploy and manage by driving greater synergy between horizon views in end-user computing and the underlying infrastructure. Horizon View Media Services for 3D graphics delivers the first phase of a virtual GPU. This enables basic 3D applications requiring DirectX9 and OpenGL in the virtual desktop without the need to purchase physical GPU or specialized client hardware. Support for Aero and Office enable a new generation of productivity applications for end users. We have to not only consider the capabilities vmware highlight, but think of customer own data-center best practices and how they want to see those applied to desktop computing.
Note: enabling this feature for desktops uses more CPU resources pre-desktop to do the graphics rendering.

Storage Access Optimization with a Content-Based Read Cache

- In-memory cache of common block reads

- Applicable to all types of desktops

- Completely transparent to the guest

- 100% server based

- No special array technology needed

- Reduced storage costs

- Reduced peak IOPS load on storage

- Ensure consistent user experience

- Handles worst-case peak usage

- Hypervisor-enabled antivirus/anti-malware

- Single AV appliance per ESX host
- Publicly announced commitments from:
  - Trend
  - Symantec
  - Kaspersky
  - Sophos
- Reduced IO load = less money
- Improved security = AV module itself is now isolated and more difficult to evade

Leverage of vShield App for Better Than Physical Network Protection

- Protect individual desktops with transparent, application-aware firewall.
- Network traffic is monitored and directed to only business appropriate systems.
- Stop malware outbreaks dead in their tracks.
- Secure sensitive applications from unauthorized access.
- Policy driven and moves with your desktop, regardless of physical networks.
- Potential for third-party value adds with DLP.
- End-to-end monitoring of infrastructure.
- Included PCoIP performance monitoring.

- Desktop, pool, and user contexts.

- Self-learning performance analytics.

- Automated alerts.

Remediation guidance

- Get to the root cause quickly; reduce MTTI.

- Respond proactively before support calls.

- Remediate quickly and accurately.

- Improve resource utilization by identifying over-provisioned hardware and track down bottlenecks.

Whether you are involved in view desktop proof of concepts, view pilots, or view plans and design activities, you know that understanding the systems, network, and storage impact of desktop user activity is critical for a successful view deployment. Therefore, it is important to simulate typical user workloads across a variety of compute tasks to accurately configure and size a view deployment.

We can prove "Smart" showback/demo/business use cases in the PoC/Pilot and creating a more compelling TCO/ROI

At the heart of a vCenter operations analytics engine is its sophisticated and highly accurate ability to understand normal behavior at the metric level. We call this *dynamic thresholding,* and we compete eight different algorithms to establish normal for a given period of time. vCenter operations algorithms do not assume any distribution of data and use nonparametric methods to determine what is normal. The algorithms consider things like cyclical patterns and calendar-driven patterns. For example, you can learn about end-of-month periods, holiday periods, etc.

The image shows an actual screenshot of vCenter Operations. This graph shows an example of how accurate our dynamic thresholding is.

- The blue line represents the actual metric behavior and you can see that this metric is a computed metric: the average hits across and entire web server farm. Note that the behavior is as expected for a web application that is primarily used during the day.

- The gray (or white) bands around the metric graph indicate the upper and lower bands of normal behavior calculated for this metric by vCenter operations. Note that in the middle of the night, the bands are very tight because there is very little variance in use of the application at night. Notice that the bands are considerably wider during the day when there is more variance in the use of the application.

- You can also see the red zone where this particular metric has breached its dynamic threshold. As discussed, in any complex IT system, there are going to be some number of metrics that exceed their thresholds.

If we were alerted to every one of them, we'd provide no more value than hard threshold-based alerts. Instead, the Operations tool does something far more sophisticated to give you alerts in the context of a growing performance problem.

Details on Analytics

The VMware Operations engine leverages many different algorithms.

In fact, every 24 hours, the vCenter Operations analytical engine pulls the full history of each metric found in the repository (FSDB). Then, the Operations engine runs that data history through eight difference algorithms (listed next), and each one determines an expected upper lower and a lower level for that metric. Once this is done, yet another algorithm (the ninth algorithm) is used to competitively score each of the results from the eight algorithms to determine which of the eight algorithms "wins" (i.e., best represents the data pattern). The effect of this process is to produce the optimal range of hour-by-hour normal behavior for the next 24 hours, or *dynamic threshold (DT)*, for each performance metric.

To further illustrate this sophistication, here is a brief description of the eight algorithms that are competitively applied:

- An algorithm that can detect linear behavior patterns (e.g., disk utilization, etc.).

- An algorithm that can detect metrics that have only two states (e.g., availability measurements).

- An algorithm that can detect metrics that have a discrete set of values, not a "range" of values, (e.g., "Number of DB User Connections," "Number of Active JMVs," etc.).

- Two different algorithms that can detect cyclical behavior patterns that are tied to calendar cycles (e.g., weekly, monthly, etc.)

- Two different algorithms that can detect general noncalendar patterns (e.g., multimodal)

- An algorithm that works, not with time-series or frequently measured values, but with sparse data (e.g., daily, weekly, monthly batch data)

There is of course a ninth algorithm that judges which of the eight should be used for the next 24-hour period.

Pragmatic Steps for Desktop Virtualization Success

The adoption journey highlighted here is based on real customer experiences to help you avoid common pitfalls and leverage best practices.

We have done the following:

- Standardized the key steps to a successful adoption

- Summarized common experiences and uncovered unique ones

- Provided feedback on what is and is not working in the real world

- Interviewed over 50 customers worldwide

The "VMware Journey" is the pragmatic approach used by VMware customers to implement virtualization with cloud computing capabilities in order to transform IT and deliver IT as a service. The Journey encompasses the knowledge and experience of hundreds of customers over the past 10 years. VMware refreshes the research behind the Journey every year to understand how new technologies and implementation approaches enable VMware customers to realize their goal of IT transformation.

Customers start by virtualizing infrastructure, development, and test environments for mission-critical applications. Virtualizing these low-governance workloads provides IT with the confidence to virtualize more sophisticated workloads later in their Journey.

As confidence, sponsorship, and the value from virtualization spread, IT shifts to focus on virtualizing mission-critical, multitier applications in production environments. This shift in focus increases the availability of all workloads and improves SLAs to the business. In summary, this is the pragmatic approach to virtualization and cloud computing.

Then, IT leverages this focus on applications to drive pervasive changes to outbound IT service levels including high-governance workloads. This focus on the business fosters better communication and partnership between IT and lines of business and results in better overall business agility.

IT Perspective

- Virtualization confidence
- Expand beyond server consolidation
- Use advanced capabilities
- Select initiatives to the cloud

Drivers Become More Business Oriented

- Reduce OPEX costs
- Improve productivity
- Security + compliance
- Improve SLA
- Faster time to market

Key Steps to Adoption

These primary elements drive adoption:

- Addressing constituents
- Staged deployment

Addressing Constituents: When changing end-user environments, there are additional constituents besides IT that must be considered for success.

IT: IT confidence drives how aggressively they will change their end-user environments and embrace newer models for end-user technology.

LOB: The business owner is driving the requirements and SLAs.

End Users: Unlike infrastructure, desktops serve a purpose of delivering productivity environments for workers, so their satisfaction is important.

Staged Deployment: Unlike server virtualization that has typically transitioned from evaluation to production quickly for the early use cases, changes to end-user environments need to be phased to minimize any disruption to the user's working environment. End-user adoption and acceptance are key to the success of any change, so taking the time to address their needs ensures a greater rate of success.

A solution assessment helps to understand if the solution does in fact address the business triggers. Additionally, a gap analysis of the current environment against the desired end state helps to define what the needs are for the transition as well as identifies any technical roadblocks to get there.

Pilot: It is important to test your prototype with real live users because a lab can never fully know how people will really behave with their systems. This also helps to engage the users in the process of refining an environment that is built for them.

Production: Once tested and tuned, this is where it is rolled out to the rest of the identified users. From here it's really about "rinse and repeat" to more users by each identified business trigger.

- **Value**: Interestingly enough, everyone had to justify value in order to get approval.

Solution Assessment

1. Identify the business trigger or what is the catalyst for change .

2. Map the trigger to the solution and develop a business case.

3. Define the sponsorship, ownership, and work team.

4. Baseline the current performance.

5. Audit the environment, users, applications, and skillset.

This stage in adoption helps to clearly map the need to the solution and helps you understand the gap between today's environment and the desired state to avoid any surprises.

Use Case: The Follow Me Desktop

Easily managed and accessible desktops for on-campus end user

Common Challenges

- M&A, expansion
- High desktop costs
- Real estate constraints
- Limited access
- Bring your own device
- Contractor/temp employees
- Painful OS migrations

Solution Benefits

- Accelerate integration
- Reduce TCO up to 50%
- Flexible access and mobility
- Secure access from any device
- Instant provisioning and deprovisioning
- Guaranteed updates and patches

The all-purpose VDI use case that addresses these challenges is associated with traditional PC deployment and management. Here is a longer list of challenges:

- Unsustainable IT operations
- High support costs
- Poor incidence resolution
- Lots of employee downtime
- High growth rate via M&A
- Aggressive market expansion
- Compliance for divesting LOBs
- Running out of office space
- Reduced real estate overhead
- Reduced power costs
- Limited user access
- No remote access
- Unable to use personal devices
- Unable to leverage new devices

Use Case: The Business Process Desktop

Delivering easily managed and integrated desktops for all agents.

Common Challenges

- Off-shoring/outsourcing business services
- High management costs
- Expensive infrastructure

- Rigid deployment

- Remote call centers

- Remote dev and test

Solution Benefits

- Secure access and control

- Reduce TCO up to 50%

- Data center cost avoidance

- Elastic resources instantly scale up and down

- Integrated UC

- Secure corporate IP in datacenter

VMware View delivers an integrated, flexible solution to manage all agent desktops more efficiently, improving business flexibility, continuity, and performance while reducing costs and increasing data security.

Common Obstacles When Getting Started

Focus
No sponsor, use case, budget, or success criteria

Team Dynamics
VDI cuts across more groups than server virtualization (network, storage, server, security, desktop, apps).

Perception
Confusion from legacy app presentation market, mudslinging from external and internal groups

Environmental Landmines
Don't know what the end-user environment looks like and if this change will work

Tips for Getting Started on the Right Foot

Focus

Engage with the CIO, gain commitment, and outline IT objectives, user population, budget, and success criteria

Team Dynamics

Engage with the teams early in the process to gain support and create new cross-functional teams

Perception

Spend time to educate key constituents on the current solution from a business and technical perspective

Environment

Assess the environment first before building the adoption plan; know the landscape and avoid landmines

Assessment Overview

- Gather baseline performance statistics

- Evaluate internal IT skill set

- Evaluate user profiles

- Identify and segment user populations by usage patterns and user experience requirements, if applicable.

What types of devices and connectivity do these users have?

- Evaluate applications

- Understand application attributes

- Identify application virtualization candidacy

- Prioritize applications for modernization

- Audit the environment (tools from Lakeside and LWL are available to help gather much of this information)

Understand environmental components like:

- Infrastructure (all the hardware like servers, storage, etc.)

- Applications

- Network

Process

Check to see if there is anything that could hold up your process in going virtual.

Pilot

1. Design and document the solution architecture prototype.

2. Define and approve project milestones.

3. Roll out to less than 200 users for controlled testing.

4. Watch for business and technical obstacles that can stall production.

5. Monitor and tune performance.

6. Identify metrics for value reporting.

In this phase you build, test, and tune a prototype until it meets the desired expectations for a production environment.

Pilot: Common Technical Obstacles

Server

Server virtualization setup is not exactly portable for end-user desktops

Image Design

Existing physical images are not exactly portable into virtual architecture

Storage

Storage architecture not compatible for end-user environments; it is about more than just terabytes

Network

Lack of visibility into configuration, available bandwidth, and planning for access

Apps

Keeping apps as they are

End Users

Poor expectation setting and not including them in definition or testing

End Users

- Need to understand what a pilot is and set their expectations that there will be issues but they are integral in making this

- Leaving them out of the process slows adoption and breeds negative perceptions

- Involve them in UAT and leverage for evangelism

Storage

- Storage architecture not compatible for end-user environments

- Plan for the right scalability and performance, not just space

Network

- Lack of understanding on current state of network

- Proactively understand the available bandwidth for performance

Pilot: Technical Best Practices

Server
Build on server virtualization skills by taking desktop virtualization PSO courses and understanding differences.

Image Design
Optimize image for virtual deployment. Offload antivirus, turn off overhead processes, stream apps instead of doing a native install.

Storage
Design storage for scalability and performance (IOPS, spindles) and not just space.

Network
LAN = quality of service. Wireless = enough access points. Remote sites = do you need to upgrade the carrier SLAs for dropped packets? Home users = determine SLAs for non-corporate-provided network.

Apps
Virtualize as many as possible and leverage methods like streaming instead of installing everything locally.

End Users
Involve them in the planning process to understand performance requirements and include in UAT.

Network: Make sure to mention LAN: QoS, and remote sites carrier SLAs for dropped packets.

Apps: All recent customers said the apps are the challenge and the ticket to freedom. If you can get them to be portable so they are easier to manage, you will have the flexibility you need to deliver them anywhere.

Pilot: Business Obstacles

Expectations
Looking for capex savings like server consolidation. Unclear scope, metrics, and success criteria lead to mismatched expectations.

Process
Status quo for IT processes to support new paradigm.

Change
Change is hard; people will either resist or rebel. Lack of information can breed negative perceptions.

End Users
Leaving them out slows adoption and breeds negative perceptions.

Pilot: Business Best Practices

Expectations
Focus on the OPEX savings and benefits to business initiative. Maintain consistent, regular communication to key constituents.

Process
Redefine the process to work with, not against, virtualization.

Change
Corporate leadership needs to be the first users. Advertise internally. Provide user incentives to accelerate change. Provide IT paths to broaden career path.

End Users

Involve them early, engage in UAT, and leverage them for evangelism.

Pilot: Monitor and Tune Performance

- Infrastructure

- Server and storage: Input Output per second (IOps), capacity

- Virtual machine

- Memory and CPU usage, performance

- Network

- Bandwidth consumption per site/users/peaks

- Apps

- Launch time, conflicts, and updates

- End users

- Logon times, responsiveness, baseline physical performance comparison

Pilot: Example of Metrics for Reporting

You can use the following as a guide to track metrics:

Track and Report OPEX Savings

Define metrics and understand baseline before transformation. Evolve metrics as you track them along the evolution of the value journey. It is not just about consolidation anymore.

Track Availability and Performance

- Business continuity is one of the major value propositions in the Business Production phase.

- Track and report both desktop/application uptime improvements and downtime risk avoidance.

End-User Productivity

Define and track end user productivity metrics impacted by the solution. This includes the number of devices adopted, decreased downtime, reduced time to help-desk resolution, increase in remote access, etc.

Production

In this phase, customers take their finalized prototype into production and transition the remaining users to the new environment.

1. Expand the pilot to the remaining users in the use case.

2. Leverage the pilot users to evangelize and educate.

3. Monitor, optimize, and tune for scale.

4. Transition the deployment to IT Operations.

5. Report the value from metrics tracking.

6. Apply the learnings to the physical environment.

7. Expand the project to address other business triggers.

Summary

In this chapter, we learned about My Workspace and the following benefits of it:

Security: It is secure by design, with enterprise-grade security. Corporate applications and data are containerized, ensuring segregation of the user's corporate IT from their personal IT. This provides IT with the control they need, without impacting the user's excellent experience.

Enhanced end-user productivity: Users like it because they work the way they want, with the devices and apps they choose, without limitations imposed by IT.

Reduced costs: The flexible, subscription-based pricing model reduces capex at setup and eliminates unnecessary opex. Standardization brings consistency and cost reduction to support services. You will get at least a 20% reduction in TCO.

Time to value: My Workspace is a preconfigured platform, and the standardized deployment services allow most deployments to complete within six months.

Increased flexibility and agility: The self-serve portal, browser-based applications, SaaS, and streaming technologies all accelerate service provisioning, while also minimizing compatibility issues.

CHAPTER 8

The Internet of Things

In this chapter, we'll discuss the IoT solution from VMware along with its ecosystem to help customers achieve their end-to-end business objective. The Internet of Things (IoT) is a system of interrelated computing devices, mechanical and digital machines, objects, animals, or people provided with unique identifiers and the ability to transfer data over a network without requiring human-to-human or human-to-computer interaction. We will go through how all the collected data is sent to a cloud infrastructure. The sensors should be connected to the cloud using various mediums of communications. These communication mediums include mobile or satellite networks, Bluetooth, WI-FI, WAN, etc. Once that data is collected and it gets to the cloud, the software performs processing on the gathered data. This process can include checking the temperature, for example, or reading on devices like AC or heaters. We will come to know how it can sometimes also be complex like identifying objects and using computer vision on video.

The IoT space is heating up, and it opens many new doors to places where federation solutions can offer unique value. IoT comes in many forms; it is not just about the smart thermostat in your home. It can be a wearable medical device that monitors a patient's vital metrics. Or it can be a smart vending machine that remembers your favorite drinks. Other examples are connected cars and even industrial applications like instrumented wind turbines and jet engines. The Internet of Things is the place where the cyber world meets the physical world.

© Ajit Pratap Kundan 2025
A. P. Kundan, *Intelligent Automation with End-User Computing Solutions*,
https://doi.org/10.1007/979-8-8688-1312-2_8

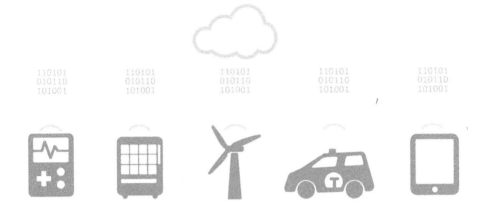

Many think the IoT is futuristic and complicated. But there are some essential building blocks. Any IoT solution requires some combination of connectivity, mobile communication, management, infrastructure, and security. Plus, of course, it needs an application lifecycle.

- Customers need an infrastructure on which they can build an IoT solution.

- Then they need to provision and manage millions of things on that infrastructure.

- Those things spit out massive data so customers also need to collect, store, and analyze data, glean meaningful insights from it, and then translate those insights into worthwhile actions.

We do this by providing the IoT essentials that enable you to realize three topline business outcomes:

- Manage millions of things as easily as managing one

- Put data from things into action

- Make IoT a business reality

How do we make these outcomes possible? Through these three ways:

- Things management

- Data capture and analysis

- Cloud-mobile service delivery

Things Management

Take the smartphone. It's a "thing" on the Internet of Things. With VMware software, we make managing millions of such things as easy as managing one. In fact, we do this today with many customers who are managing fleets of things, such as sensors and cars. For instance, we've been working with a well-known beverage company who deploys VMware Workspace ONE and vRealize Operations to manage their smart vending machines. Smartphones are also the end user's primary IoT connection point: think printers, retail point-of-sale transacted on iPads, your Fitbit. Smartphone operating systems are a primary component in the IoT ecosystem, enabling communication between smart things like wearables and those ubiquitous smartphones. Through VMware, we integrate the OSs of all these things, including QNX and Android.

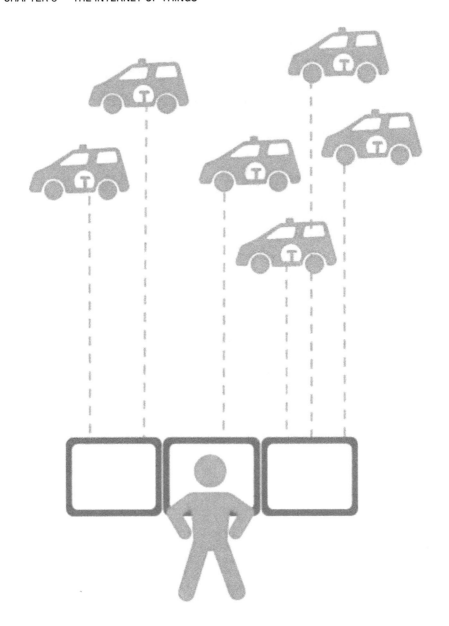

Gartner has named VMware as a leader in the latest Magic Quadrant for Enterprise Mobility Management Suites. Out of the 12 vendors evaluated, Gartner positioned VMware highest on the "ability to execute" axis. On one end of the Internet of Things are the things themselves. On the other end, you have users interacting with those things or the data from those things. Before reaching end users, that "things" data takes a journey through a processing pipeline that includes data cleanup, crunching by online analytical models, and consumption and presentation by custom applications.

Data Capture and Analysis

IoT things, including smartphones, generate a ton of data and send it back through what's called an *IoT gateway*. Gateways then send that data back to the data center. Data could stream continuously, to be analyzed and acted on later. But we can make things smarter by scanning collected data at the gateway and identifying patterns that might require action.

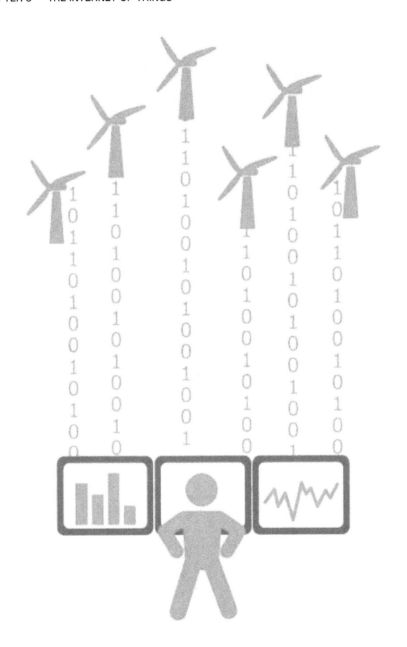

Cloud-Mobile Service Delivery

With the combination of VMware Workspace ONE, vRealize Operations, and Log Insight software, customers can capture massive amounts of "thing" data, run dynamic thresh-holding, analyze operational anomalies, and trigger mission-critical actions. For example, you could alert the operator if an offshore oil pump is overheating, before the entire system halts. Or replace the batteries on a smartphone used for critical home-healthcare. What's more, with the help of technologies such as vSphere Big Data Extensions, the Pivotal Big Data Suite, and the Federation Business Data Lake, customers can develop advanced Big Data Analytics apps to drive sophisticated business analyses and insights. Additionally, running in the datacenter, Pivotal Spring XD integrates with a host of APIs and data sources, enabling data transformation, including real-time event processing, through the data-processing pipeline.

Cloud-mobile service delivery encompasses the design, development, deployment, and operation of IoT custom applications. To enable end-to-end IoT service delivery, customers need a cloud-mobile platform where custom apps run. It's critical that that be an "opinionated platform" (versus "vanilla stack") so you can develop and operationalize apps easily and at scale.

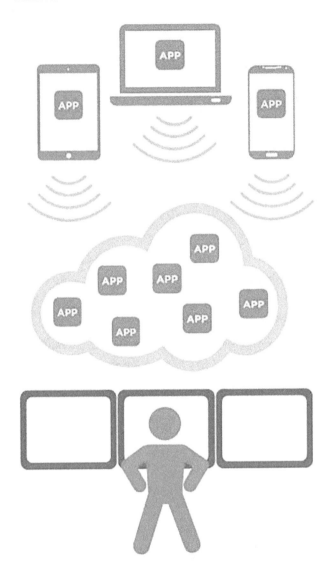

The Federation offers the platform solutions customers need for IoT service delivery, whether from a private software-defined data center, on the enterprise hybrid cloud, in the vCloud Air public cloud, or on the Pivotal Cloud Foundry application platform. For example, Pivotal Cloud Foundry takes care of all operations; a developer simply runs one command on their app to get it up and running (e.g., CF push or CF scale). Similarly, Pivotal Spring Cloud (which sits on the PCF platform) lets developers quickly and easily build cloud-native architectures with Java: one Java annotation and you have a fully integrated service system.

These platforms, in combination with the management tools Workspace ONE and vROps, take the complexity out of IoT. And with less complexity to deal with, customers don't feel as overwhelmed or confused when they think about the IoT. They're free to start thinking about IoT service delivery for their customers.

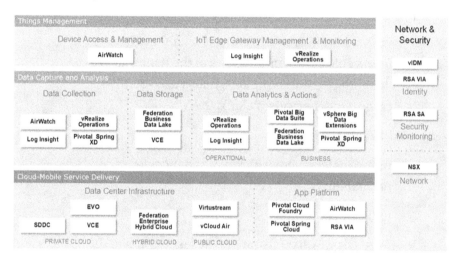

Here's the big-picture view of how Federation solutions play together to enable things management, data capture and analysis, and cloud-mobile service delivery:

Smart Vending Machine

- Workspace ONE collects beverage usage data and manages machine maintenance and supplies.

- It fetches real-time data from beverage dispensers (status, number of drinks served, available supplies, etc.).

243

IoT Edge Gateway

- vROps can collect data from Intel gateways, organize data views, analyze incoming data streams, issue user defined commands, and raise alerts.

- Adapter and gateway agents are under development.

- It gives the vCloud Air account a vRealize Operations Manager instance.

Connected Car

- vROps running on VCA collects telemetric data from vehicles, organizes views of the data, analyzes incoming real-time streams, issues user-defined commands, and raises alerts.

Industrial Turbine

- Improves jet engine efficiency and increase service profitability

- In production

- Together we offer customers the fastest path to the IoT so they can take full advantage of best-of-breed converged infrastructure and hyper-converged infrastructure where the software and hardware solutions are engineered and tested to work optimally together, offering you choice and flexibility, without complexity.

Why IoT?

Organizations in a *variety of industries* are using the IoT to operate more efficiently, better understand customers to deliver enhanced customer service, improve decision-making, and increase the value of the business.

- An IoT ecosystem consists of web-enabled smart devices that use embedded processors, sensors, and communication hardware to collect, send, and act on data they acquire from their environments.

- IoT devices share the sensor data they collect by connecting to an IoT gateway or other edge device where data is either sent to the cloud to be analyzed or analyzed locally.

Top 10 Strategic IoT Technologies and Trends

Trend No. 1: Artificial Intelligence (AI): Data is the fuel that powers the IoT, and the organization's ability to derive meaning from it will define their long-term success.

Trend No. 2: Social, Legal, and Ethical IoT: These include ownership of data and the deductions made from it, algorithmic bias, privacy, and compliance with regulations such as the General Data Protection Regulation. "Successful deployment of an IoT solution demands that it's not just technically effective but also socially acceptable."

Trend No. 3: Infonomics and Data Broking: The theory of infonomics takes monetization of data further by seeing it as a strategic business asset to be recorded in the company accounts. The buying and selling of IoT data will no longer be experimental by 2028 and it will be essential.

Trend No. 4: The Shift from Intelligent Edge to Intelligent Mesh: The shift from centralized and cloud to edge architectures is well underway in the IoT space. These mesh architectures will enable more flexible, intelligent, and responsive IoT systems, often at the cost of additional complexities.

Trend No. 5: IoT Governance: As the IoT continues to expand, the need for a governance framework that ensures appropriate behavior in the creation, storage, use, and deletion of information related to IoT projects will become increasingly important.

Trend No. 6: Sensor Innovation: The sensor market will evolve continuously through at least 2030, as advancements in materials, miniaturization, and edge AI unlock new capabilities and cost efficiencies. New classes of sensors will enable detection of a much broader range of situations, behaviors, and environmental conditions by 2027.

Trend No. 7: Trusted Hardware and Operating System: "We expect to see the deployment of hardware and software combinations that together create more trustworthy and secure IoT systems...."

Trend 8: Novel IoT User Experiences: User experience is driven by four factors: new sensors, new algorithms, new experience architectures and context, and socially aware experiences.

Trend No. 9: Silicon Chip Innovation: By 2023, it's expected that new special-purpose chips will reduce the power consumption required to run IoT devices.

Trend No. 10: New Wireless Networking Technologies for IoT: IoT networking involves balancing a set of competing requirements. In particular they should explore 5G, the forthcoming generation of low-earth orbit satellites, and backscatter networks.

Benefits of IoT

IoT offers a number of benefits to organizations, enabling them to:

- Monitor their overall business processes

- Improve the customer experience

- Save time and money

- Enhance employee productivity

- Integrate and adapt business models

- Make better business decisions

- Generate more revenue

The IoT and Public Safety

The Internet of Things is quickly transforming almost every industry, and public safety is no exception. Technologies used to accomplish smart facilities, smart cities, smart vehicles, and the like are combined to provide critical data for first responders and agencies at all levels of government.

Some examples include:

- Drones used by law enforcement or wildfire personnel to gain visibility in areas that are remote or unsafe

- Highway sensors and adaptive traffic management technologies used to provide first responders with faster routes to emergency locations

- Sensors to track law enforcement weapons and other equipment

- Cameras, sensors, and connected microphones used in public spaces for real-time gunshot monitoring and to help triangulate the location of a shooter

- Smart vehicles that receive routes and other instructions from emergency systems

Regardless of the use case, the hardware deployed to collect and process data must be monitored and managed to ensure device health and to secure this new IoT infrastructure.

Real-Time Monitoring

For mission-critical systems, real-time monitoring of the infrastructure is essential. Administrators require tools to understand the health of remote, unmanned devices within an IoT use case and accurate alerting to detect anomalies that may lead to device failure.

Device and Software Lifecycle Management

Sensors, cameras, and other devices used for public safety likely have firmware that needs to be updated on a regular basis, especially if security vulnerabilities are discovered. In addition, the edge systems and gateways that aggregate telemetry from the connected "things" will need operating system updates, security patches, and updates for installed applications. There must be a method to deploy software and firmware updates to devices over-the-air (OTA) since most public safety use cases require devices to be deployed across long distances or in locations difficult to reach, such as in walls or ceilings. It is not scalable to physically visit each device when an update is required.

Three-Tier Architecture

More and more often gateways are added to the IoT architecture. These devices serve as the decoupling and control point between sensitive IoT devices and the cloud or data center. In addition, IoT gateways are used to connect "simple" devices with no operating systems, like sensors, to the network. This results in a three-tier architecture: simple device to gateway to data center. Therefore, the edge systems or gateways need to be managed along with the connected devices. This parent-child relationship (simple device and gateway) needs to be clear in the management solution so that the administrator understands which gateways are proxying the management of which connected things.

Managing Heterogeneous "Things"

Because different use cases require different devices, it is common for an agency to have diverse IoT devices from varying manufacturers with different operating systems and communication protocols. For this reason, the management system must be flexible enough to support various devices.

In the world of IoT, we are no longer talking about hundreds or even thousands of connected things. We are talking about millions of sensors, cameras, robots, vehicles, etc. A management system is required that can not only support this scale but provide tools to manage millions of devices as easily as one.

Introducing VMware Pulse IoT Center

VMware Pulse IoT Center is an end–to-end management application that helps OT (operational technology) and IT (information technology) onboard, manage, monitor, and secure all IoT devices throughout their lifecycle, regardless of device type or manufacturer. This solution helps administrators onboard edge systems (e.g., IoT gateways) and their connected devices (e.g., sensors, cameras, etc.), collect telemetry from them, analyze the data stream to determine the operational health and status, and maintain these devices with timely software, OS, and firmware updates. The solution also detects operational anomalies, takes action on them in real time; enables pushing files, applications, and configurations to the devices and systems; and manages the device lifecycle.

VMware Pulse IoT Center utilizes an agent that sits on the IoT edge systems and gateways. A critical component of this agent is LIOTA: Little IoT Agent. Liota is an open-source project offering convenience for IoT solution developers to create edge system data orchestration applications. Liota has been generalized to allow, via modules, interaction with any data-center component, over any transport, and for any IoT edge system. This SDK/framework provides a method for writing portable, modular, dynamic IoT gateway data orchestration apps. Liota applications can be written to collect telemetry from any "things" connected to edge systems, regardless of manufacturer, device type, or communication protocol. Please visit GitHub (*https://github.com/vmware/liota*) or more information on Liota.

VMware Pulse IoT Center is licensed per managed object (MO), which includes edge systems, IoT gateways, and the connected things. Devices with the Pulse agent installed, i.e., edge systems/gateways, is a higher unit price than simple connected devices. Tiered pricing may apply to large deployments.

Many organizations are realizing that the data center is going to the edge, often driven by the need for real-time analytics close to the data source, either due to limited bandwidth or due to data privacy and security. For example, temporary central command centers during an emergency require many systems and compute in remote locations. VMware recognizes this need and is developing a hyper-converged appliance that includes both Pulse IoT Center and analytics from industry leading third parties, all preconfigured on top of VMware Cloud Foundation. Project Fire could be considered a "pop-up data center" to quickly deploy an IoT use case in hours or days versus weeks or months.

Data collected in public safety use cases, such as criminal information or locations of citizens, is often sensitive and subject to multiple regulations. In addition, the equipment used by law enforcement and emergency response teams can often be life-threatening if misused. Therefore, the security of the IoT architecture must be the highest priority when designing a solution. This is why VMware is exploring integrations between Pulse IoT Center and VMware's existing offering for network virtualization and security (NSX) to provide security at the edge of the network.

The VMware Partner Ecosystem

Implementing a complete IoT solution is a huge challenge for public agencies, as they often need to cobble together offerings from various vendors with few standards and little guidance.

According to Gartner, the rate of failure in IoT projects is expected to double by 2027, as existing technologies and architectures struggle to meet the scalability, complexity, and integration demands of large-scale IoT deployments.

VMware and its rich partner ecosystem can:

- Take the guesswork out of IoT implementation decisions to help customers address complex IoT use cases across multiple use cases

- Help bridge the gap between the IT and OT (operations technology) worlds

- Help deploy an end to-end IoT solution from the edge to the data center

Thing/OEM Manufacturers

Hardware considerations are critical to the success of an IoT deployment. For example, it is important to make the distinction between consumer-grade and enterprise-ready devices. Many sensors, cameras, and other devices may not be manageable remotely or have the security features IT requires. In addition, some use cases will require solar-powered edge systems or hardware specifically made for rugged, outdoor, or refrigerated environments since devices often reside in remote areas.

Examples include Dell, Axis, ABB, Zebra, HPE, Intwine, V5 Systems, Harman, Denso, Bossa Nova, Keiback, and Peter.

How the Customer Benefits

- Deploy connected things that are IT approved and prevalidated

- Have a solution that can manage and configure many heterogeneous things in a single console

- Customer of embedded systems gets the benefit of getting curated industry-specific solutions that are "manageable" out of the box

GATEWAY/EDGE SYSTEM MANUFACTURER

With the advent of the three-tier architecture, gateways have become a major component of any IoT-ready architecture. These gateways do a lot of the heavy lifting for the things connected to them like storage and edge analytics, such as integrating protocols for networking and facilitating data orchestration securely between edge devices and the cloud.

VMware has partnered with gateway manufacturers that cater to specific use cases as well as broad ones to provide a great selection for our customers.

Examples include Dell, ADLINK, Eurotech, Samsung, Hitachi, Cloudsense, HP, ITG, and Axis Communications.

How the Customer Benefits

- The customer gets the edge system/gateway of their choice along with the ability to do software and firmware lifecycle management at scale.

- The solution provides a gateway/edge system that is "manageable" and IT approved.

- The solution provides more efficient and secure gateways (with virtualization).

- The customer enjoys a multivendor, heterogeneous gateway architecture that can all be managed, monitored, and secured from a single pane of glass.

Business Applications and Analytics

Applications, and in particular tools for analyzing and extracting value from the data generated by connected things, are essential for maximizing the benefits of any IoT implementation. VMware has partnered with leading global players whose expertise spans the gamut of industry sectors.

Examples include VizExplorer, SAP, IBM, CenturyLink, and Zingbox.

How the Customer Benefits

- The customer gets a pre-integrated IoT application solution along with the underlying IoT infrastructure management and monitoring capabilities.

- The customer gets the flexibility of choosing on-premises capabilities as well as cloud capabilities for their architecture.

IoT Platform Vendors

At the heart of IoT infrastructure, the IoT platforms play a crucial role ensuring the seamless integration of different IoT hardware devices and IoT applications that support analysis, data visualization, etc. VMware has partnered with global companies whose IoT platforms are market leading across key industry verticals. Together, they will offer IoT starter kits that are industry and use case specific and sometimes on-prem. For example, they will be delivered directly to the factory floor, in a hospital, or on an oil rig or power plant to significantly reduce the time it takes for customers to deploy IoT. These packaged solutions will include VMware IoT solutions to manage, monitor, and secure a variety of IoT edge systems and connected devices as an integrated part of the IoT platform. It will also include VMware's hyper-converged infrastructure software to save time, effort, and cost to deploy IoT.

Examples include ThingWorx, Fujitsu, SAP, IBM, and Bosch.

How the Customer Benefits

- Get operational analytics as well as business analytics in the same solution

- Have the option of a completely integrated (software+hardware) mini data center that can be deployed at the edge

Systems Integrators

Global leaders in systems integration will have a critical role in the design and implementation of many IoT projects. These projects will require expertise across a wide range of information and communications technologies and the ability to execute at scale and, in many cases, across multiple locations. VMware has partnered with a select group of market leaders that will embed VMware's IoT solutions into their recommended architectures.

Examples include Deloitte, DXC, Tech Mahindra, Cognizant, Wipro, Atos, and Fujitsu.

How the Customer Benefits

- The customer gains a pre-integrated and pre-validated IoT solution that meets the requirements of both OT and IT organizations.

- The customer gains the ability of on-premises capabilities.

- The customer gains the expertise of industry leading IT providers (VMware) and industry specific knowledge (System Integrator partners) in setting up your IoT architecture.

- The customer gets a solution that is purpose built for their use case and can be managed from edge all the way to the cloud.

IoT and Its Alignment with AI and Edge Computing

Convergence of IoT, AI, and Edge Computing

- **IoT**: The Internet of Things generates massive amounts of data from connected devices, sensors, and systems.

- **AI**: Artificial intelligence processes and analyzes this data to extract actionable insights, automate decision-making, and enable predictive capabilities.

- **Edge Computing**: Edge computing brings computation and data processing closer to the data source, reducing latency, reducing bandwidth usage, and improving response times for IoT applications.

Key Areas of Alignment:

- **Real-time data processing**: AI models running on edge devices process IoT data in real time, which is essential for time-sensitive applications like autonomous vehicles, industrial automation, and healthcare monitoring.

- **Scalability**: Edge computing offloads computational tasks from centralized cloud systems, enabling IoT systems to scale efficiently while maintaining performance.

- **Energy efficiency**: By minimizing data transmission and optimizing resource use, edge computing makes AI-powered IoT systems more energy-efficient.

- **Enhanced security**: Data processed locally on edge devices reduces the risk of breaches during cloud transmission, aligning with the growing need for secure IoT ecosystems.

Potential Future Applications and Extensions

1. **Smart Cities**

   - **AI-driven traffic management**: Integrating IoT sensors with AI algorithms at the edge for real-time traffic flow optimization.

   - **Predictive maintenance**: Sensors in infrastructure (e.g., bridges, roads) send data for AI-driven failure predictions.

   - **Energy optimization**: IoT-enabled smart grids use AI at the edge to dynamically adjust energy distribution.

2. **Healthcare**

- **Wearable IoT devices**: AI processes patient data locally on wearables to deliver real-time health insights, detect anomalies, and notify caregivers.

- **Remote surgery**: Low-latency IoT systems powered by edge computing enable real-time control and feedback for tele-surgery.

3. **Industrial IoT (IIoT)**

- **Autonomous manufacturing**: AI at the edge analyzes IoT data from machinery for automated quality control and fault detection.

- **Supply chain optimization**: IoT sensors paired with AI algorithms track goods and optimize logistics dynamically.

4. **Agriculture**

- **Precision farming**: IoT sensors and AI at the edge analyze soil conditions, weather data, and crop health to optimize water usage, fertilizers, and harvesting.

- **Autonomous machinery**: Edge-enabled AI powers drones and robots for planting, monitoring, and harvesting.

5. **Retail**

- **Smart shelves**: IoT devices track inventory in real time, while edge-based AI predicts restocking needs and customer demand.

- **Personalized shopping**: AI processes IoT data locally in stores to provide customers with tailored recommendations.

6. **Connected Vehicles**

   - **Autonomous driving**: Edge AI processes sensor data in real-time to make split-second decisions.

   - **Vehicle-to-everything (V2X) communication**: IoT devices in cars exchange data with infrastructure and other vehicles, enhancing traffic safety and efficiency.

7. **Environmental Monitoring**

   - **Disaster prediction**: AI models analyze IoT sensor data to forecast natural disasters like floods, earthquakes, and wildfires.

   - **Sustainability tracking**: IoT-enabled systems monitor environmental parameters like air and water quality, processed in real time via edge AI.

Key Future Challenges and Opportunities

- **Standardization**: Developing universal standards for IoT and edge AI integration

- **AI model optimization**: Enhancing AI algorithms for lightweight operation on edge devices

- **Interoperability**: Ensuring seamless communication between diverse IoT and edge systems

- **Ethical AI and Privacy**: Addressing ethical concerns in IoT data processing, particularly at the edge

Summary

The VMware IoT solution along with its ecosystem vendors help organizations to simplify IoT complexity. We have gone through the three steps of managing things, collecting data, and sending this data to different cloud types for data analytics. The solution simplifies the end-user experience and reduces IT costs by combining all three stages. It can also help in securely delivering these reports to any device. End users gain freedom of mobility through anytime, anywhere access. The results are simpler, centralized, policy-based management and control of IT consumerization for the administrator. IoT, coupled with AI and edge computing, is set to redefine industries by enabling hyper-efficient, secure, and intelligent systems, laying the foundation for transformative innovations in the years ahead.

CHAPTER 9

Delivering the Digital Workspace

In this chapters, we'll cover how users should have one place to securely access all the applications, files, social tools, and online services they need, from any device they choose, from anywhere. We also look at how to enable the software-defined workspace and allow people to work at the speed of life as it is about end users and how they want to work, whether they are in the coffee shop, at home, on the road, or in the office accessing their applications and data that they need. We will also cover how customers can take advantage of the full capabilities up and down the stack that VMware brings to bear around the software-defined data center and extend these capabilities to end users through the workspace.

To make a digital transformation, two things are paramount for success:

- Consumer simplicity
- Enterprise security

© Ajit Pratap Kundan 2025
A. P. Kundan, *Intelligent Automation with End-User Computing Solutions,*
https://doi.org/10.1007/979-8-8688-1312-2_9

Enterprise companies tend to get the security part right but often have issues with consumer simplicity. Both are crucial steps in driving the digital workplace. The entire world is going digital. Processes are shifting from:

- Analog to digital

- Paper to mobile

- Manual to automated

Social computing takes on a whole new paradigm, where it's becoming much more dynamic:

- Initially, social computing was thought of as one person communicating with another person.

- Now, social has shifted to include:

    - The person's communication with the machine

    - The machine's communication with the machine

    - The machine's communication back to the person

That's how we think about the transformation of the digital workspace. To thrive in this digital world, VMware has engineered a solution called the VMware Workspace Suite. VMware was one of the first companies to coin the term *workspace* when it acquired a small company called Propero.

With today's digital workspace framework, VMware is the only company that allows you to bring any application to any device. Any application, such as:

- Windows applications

- Web applications

- HTML5 applications

- Mobile applications

can now be experienced on any device, such as:

- Laptops

- Desktops

- Tablets

- Phones

- Vending machines, etc.

To make this possible, VMware has engineered the Workspace Suite to continuously innovate under these three pillars:

- Desktop Portfolio

- Mobile Portfolio

- Content Collaboration Portfolio

VMware has continued its innovation by linking these portfolios together under a single sign-on called VMware Identity Manager.

- This enhances the success of our digital workspace transformation by bridging customer simplicity and IT security.

- Identity Manager brings together the users' desire for choice, with IT's desire for control.

Self-Service Access

- Data security: Apps that are declined or denied access

Conditional Access

Virtual Desktops and App Publishing from One Platform

- Online or offline desktop
- Unified workspace catalog for end users

- Rich user experience for 3D, multimedia and real-time collaboration

- Secure containerized desktops for laptop/ disconnected users

- Real-time app delivery and lifecycle management

- Lowest TCO with hyper-converged and optimized SDDC

Horizon Makes Desktop and App Management Easy

Horizon centralizes end users' desktops and applications in the data center so IT can efficiently provision new clients, centralize desktop management, and improve security and compliance. It is based on seven key pillars.

Desktops and Apps From a Single Platform	Great User Experience	Just-in-Time Desktops	Smart Policies	Complete Environment Management	SDDC Integration	Flexible and Hybrid Delivery

- Best user experience

- Simplified, consistent management, no patch maintenance window

- Provisioning on-demand

- Space efficient

VMware Horizon 7 Delivers Simplified Management and Automation

Configuration/Provisioning

- View Composer single image management

- Nvidia GRID profile settings in vCenter

- Automated provisioning of 3D Desktops to GRID VGPU hosts in a cluster

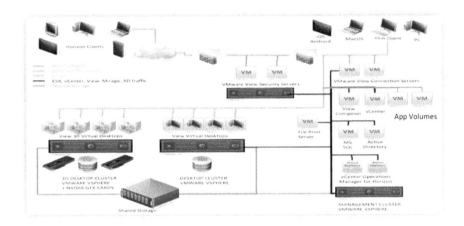

Monitoring

- Proactive end-to-end visibility with vRealize Operations Manager for Horizon

Real-Time App Delivery

- App lifecycle management via App Volumes

- Secure remote access via VMware Horizon View Security Servers

- Nvidia GRID profile and VM settings with vCenter

- Build automated pools of persistent or nonpersistent desktops

- Automated provisioning of 3D Desktops to GRID VGPU hosts in a cluster

- View Composer single image management and storage optimization

- Streamlined app lifecycle management via App Volumes

- Proactive end-to-end monitoring with vRealize Operations Manager for Horizon

Benefits:

- Simplify app deployment and updates

- Deliver, update, and retire any set of applications in seconds with App Volumes

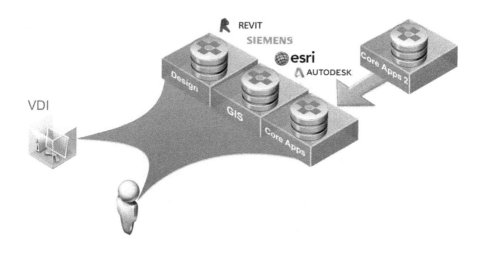

- Logically manage applications based on line-of-business requirements.

- Update immediately, or on next login

- Simplify management of all of your core applications through one AppStack

Blast Extreme

- Scale and density with GPU offload

- High performance over public networks

- Designed for any low-cost device

- Another protocol option in Horizon

- Feature and performance parity with PCoIP

- PCoIP _not_ going away

- Zero clients

- H.264 codec support

- Most devices have H.264 hardware decode support

- Hardware H.264 encode with Nvidia GRID

- Proprietary JPG/PNG codec support

- Same as used in Blast Extreme HTML/Linux

- Supports both TCP (the default) and UDP

- Native Horizon Client 4.x required

VMware Horizon: Desktop Transformation Solution for Every User

- The Workspace Suite brings together the complete Horizon desktop (AirWatch mobile platform, and content management with a common set of services underneath it, such as identity, social, catalog, and gateway services) into a complete and integrated solution.

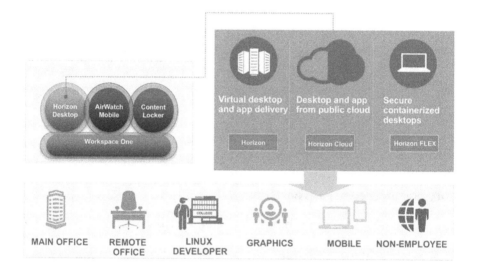

Desktop Layering Vision

- Moving from monolithic to layered desktop
- Decompose desktop into separate manageable entities

Reduced number of configurations:

Old model: Each combination of application set, base OS, and VM model required a different pool

New model: Elements dynamically combined

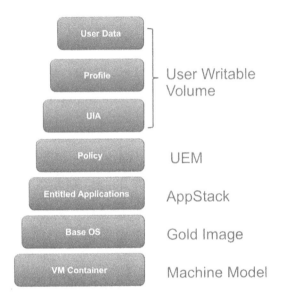

More customized user experience:

- Users get the combination of elements they need

- Not "one size fits all"

More easily managed:

- Replication of just user data, not entire VM

- Easier to re-create in a different location

- Enable DR, cloud bursting, roaming

Why Choose VMware for High-Performance Graphics VDI?

- Proven, industry-leading platform, fully integrated end-to-end solution

- Deliver the most graphically intensive applications

- Deliver a superior experience with PCoIP and Blast Extreme

- Choice of GPU technologies

- vGPU, vDGA, and others

Desktop and App Virtualization Enables Business Mobility

Desktop virtualization with a server-side GPU improves the delivery of graphics. But even with GPU-acceleration for VDI, challenges still exist:

- Traditional approaches have forced a trade-off of extremes.

- On one hand, you could deploy a GPU that is shared across many users, which is very cost-effective but offers poor performance for anyone except for maybe task or knowledge workers who have a light 3D requirement.

- On the other hand, you could deploy a GPU that's dedicated to one user at a time. We call this GPU *passthrough* or dedicated graphics. It offers great performance, but because it supports only one user, it's not scalable, and it's very expensive from a cost-per-user perspective.

- The marketplace has been missing a viable solution that addresses these trade-off of extremes.

- This is where VMware Horizon with Nvidia GRID vGPU comes in.

VMware Horizon with Nvidia GRID vGPU

- **Immersive graphics from the cloud:** Shared, workstation-class graphics accessible on any device

- **Single platform, lower costs:** VMware solution from endpoint to data center, with lower opex

- **ISV compatibility:** With a growing ecosystem of the leading 3D apps

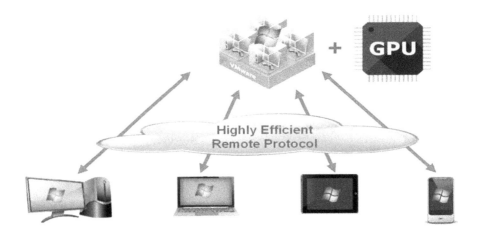

Users remote access from various devices

Why Is VMware Horizon with Nvidia GRID vGPU Better for Your Customers?

First, it allows them to cost effectively share the power of GPUs across multiple users, while retaining workstation-class graphics, without being tethered to the traditional workstation. Second, because it's built on an end-to-end VMware platform, using the industry's most successful hypervisor platform, vSphere, and the software-defined data center, so it dramatically lowers opex. Finally, VMware and Nvidia have fully certified this platform with a growing list of the most important 3D application ISVs in the marketplace so customers can deploy with confidence.

Delivering a Complete Portfolio, for Every Worker, in Every Use Case

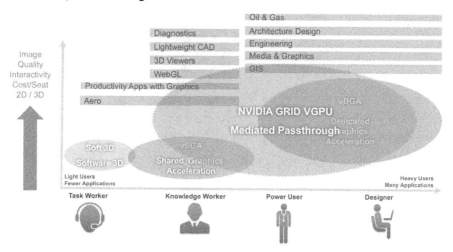

273

VMware Horizon with Nvidia GRID vGPU

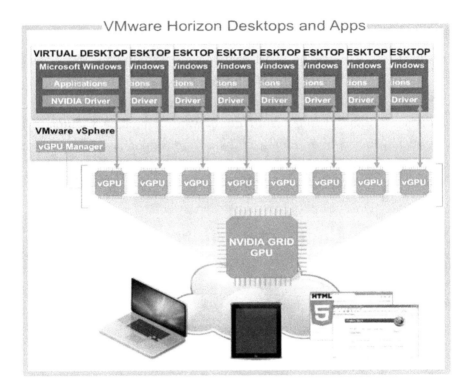

Improved Shared 3D Graphics with Multiple VMs: vSGA

- vSphere adds AMD/ATI graphics cards.

- Supports select Nvidia & AMD/ATI graphics cards.

- Enables shared access to physical 3D graphics cards for high-performance graphical workloads.

- Desktops see abstracted VMware SVGA device for maximum virtual machine compatibility and portability.

- Share single 3D graphics card with multiple virtual machines.

Benefits

- Enables truly high-performance graphics

- Cost effective with multiple VMs sharing single 3D graphics card

- Full compatibility with vMotion, DRS for hosts lacking physical 3D graphics cards

Deliver Workstation Class 3D Graphics with VDI: vDGA

This offers a full workstation-class user experience with a dedicated Nvidia graphics card.

Overview

- Enables dedicated access to physical GPU hardware for 3D and high-performance compute workloads

- Uses native Nvidia drivers

- CUDA and OpenCL compute APIs supported

- Best for super high-performance needs like design, manufacturing, oil and gas

Benefits

- Complements vSGA cost/performance

- True workstation replacement option

- Full capabilities of physical Nvidia GPUs

- High-performance compute GPU option

Troubleshooting

Problem	What to Check
vDGA VMs fail to start	If the VM has more than 2 GB virtual memory configured, check in the VMX file if the pcihole.start vmx setting is correct (installation steps).
Nvidia driver fails to install	Open up the Device manager, select "Display adapters," and see if the Nvidia card is not present or marked red. This means that the passthrough was not configured correctly. Go through the passthrough steps again (installation steps).
vSphere console has a display and can be accessed after the first view connection.	The vSphere console should no longer show up any display after the first view connection It shows either a black screen or a frozen Windows startup display. Run ontereyEnable.exe as instructed in the installation steps and reboot the VM.

Drivers for MMR

High-Fidelity User Experience

- Native desktop experience expected

- Required across task/knowledge and power users

- Locally encoded video provides a better experience

Reduced Bandwidth/Higher Consolidation

- Host-side-rendered video typically requires more bandwidth

- Host-side-rendered video cannot be cached today

- Host-side-rendered video is more CPU intensive

Branch Deployments

- Better user experience over WAN

- Lower bandwidth

- Support for branch caching/data dedup

Windows Multimedia Redirection

- VMware Horizon View introduces initial support for MMR with Windows VM desktops using PCoIP.

- With Windows MMR, only the video is redirected from the guest to the client endpoint. .

- The audio is sent within PCoIP down to the client endpoint.

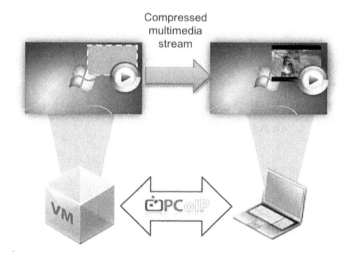

Client Requirements

- Windows MMR will work only when using the View client with Windows 10/11 clients.

- The Windows client machine must have a DXVA-capable GPU.

- Windows MMR leverages the DXVA GPU to locally render the video.

- Without a DXVA capable GPU, MMR will not function, and video playback falls back to host-side rendering.

ATI/AMD GPUs are not supported at this time.

Useful Tools

DXVA Checker: Use to check the capabilities of the GPU on the client machine.

Gspot: Use to verify video codec that is used to encode a multimedia file. It also provides native bit rate information.

Networx: Use to measure bandwidth inbound and outbound.

How to verify that Windows MMR is installed: Make sure that the administrator has selected Windows MMR during installation of the Remote Experience Agent (REA).

How to Verify that Windows 7 MMR Is Enabled

Check on the view connection server and verify that MMR is enabled for desktops.

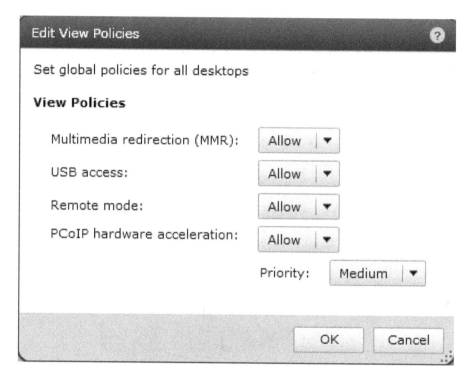

How to Verify That 3D Is Enabled on the Desktop

Check on the view connection server and verify that 3D is enabled for the desktop.

Check the Following Registry Settings

- Open RegEdit and navigate to the following folder:

    – HKEY_LOCAL_MACHINE ➤ SOFTWARE ➤ Wow6432Node ➤ VMware, Inc. ➤ VMware VVA

    – For 32-bit bit systems, make sure Wow6432 is not present.

- Make sure that the "enabled" field is set to 1.

- Make sure that the "whitelist" field has "wmplayer.exe" listed.

Real-Time Audio-Video (Webcam and Microphone) Improved Webcam Experience and Performance

- Optimized delivery of webcam and microphone traffic for View desktops

- Encoded and compressed video reduces upstream bandwidth for webcam traffic to as low as 300kbps

- Improves installation and administration of webcam devices

- Windows and Linux

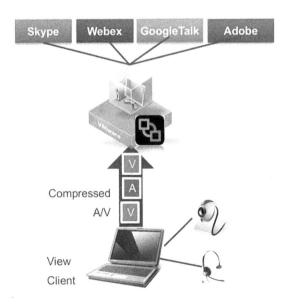

Benefits

- Improved end user experience

- Lower (100x) bandwidth consumption

Overview of Solution

- Virtual devices on the agent present the feed (audio+video) to third-party applications (e.g., Skype)

- Application-level redirection also allows the view to support a broader range of devices providing the same functionality (USB and non-USB), e.g., analog microphones and also USB microphones

- Allows use of codecs, which results in relatively better performance using fewer resources (Theora and Speex codecs)

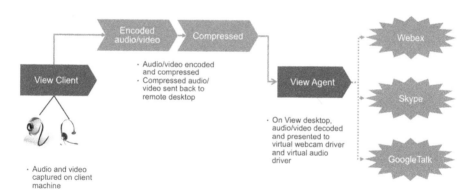

Comparing vSGA, Nvidia GRID vGPU, and vDGA

Attributes	vSGA	NVIDIA GRID VGPU	vDGA
Sharing	GPUs shared between users	GPUs shared between users	GPU dedicated to one user
Consolidation	Good consolidation for low-end graphics use cases	Good consolidation ratio (up to 8:1)	Only one user per GPU (1:1)
Performance	Lightweight rich graphics WITHOUT video acceleration	Scalable performance from entry level to high end	High end workstation performance
App Compatibility	Drivers NOT application certified	Fully certified application drivers	Fully certified application drivers
DirectX APIs	DirectX 9	DirectX 9, 10 ,11	DirectX 9, 10 ,11
OpenGL APIs	OpenGL 2.1	OpenGL 2.1, 3.x, 4.x	OpenGL 2.1, 3.x, 4.x
General Purpose Compute	Does NOT support compute CUDA, OpenCL	Does NOT support compute CUDA, OpenCL	Compute APIs with CUDA, OpenCL
Automated Management	Yes	Yes	No
vMotion / HA	Yes	No	No

Introducing MAXWELL

	M60	M6
GPU	Dual High-end Maxwell	Single High-end
CUDA Cores	4096	1536
Viewperf 12*	62 x 2	54
Memory Size	16 GB GDDR5	8 GB GDDR5
H.264 1080p30 streams	36	18
GRID vGPU CCU	2 / 4 / 8 / 16 / 32	1 / 2 / 4 / 8 / 16
Form Factor	PCIe 3.0 Dual Slot	MXM
Power	240W / 300W (225W opt)	100W (75W opt)
Thermal	active / passive	bare board

Webcam Redirection System Architecture

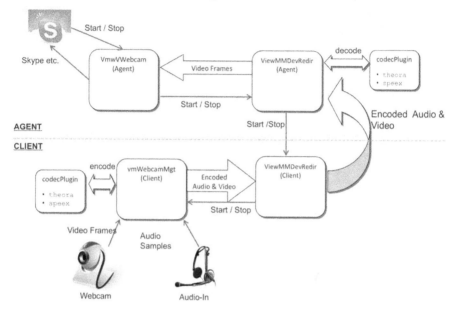

Flash URL Redirection: Streaming of Live Video Events from Adobe Media Server

- Stream live video events optimally to Horizon View desktops

- Support for live video streaming on Adobe Media Server

- Supported with Windows and Linux thin clients

Benefits

- Stream live video events to virtual desktops without affecting the data center server and network

- Enables new multimedia use cases with virtual desktops

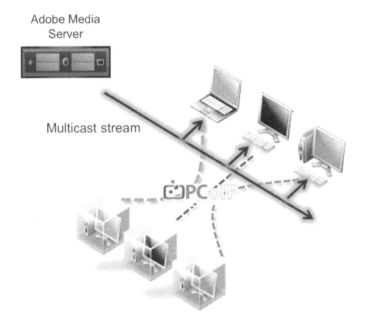

How Flash URL Redirection Works

The flash URL redirection feature uses JavaScript that is embedded inside an HTML web page by the web page administrator. Whenever a virtual desktop user clicks the designated URL link on a web page, the JavaScript intercepts and redirects the SWF file from the virtual desktop session to the client endpoint. The endpoint then opens a local Flash Projector outside of the virtual desktop session and plays the media stream locally.

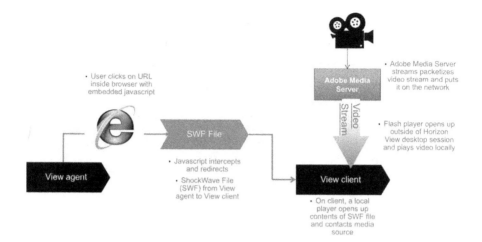

Configuration
Agent

- Verify that the Remote Experience Agent with the
  Flash URL Redirection option is installed on the virtual
  desktop. To verify:

    - Start a virtual desktop session that uses PCoIP. RDP
      is not supported.

    - Verify that the ViewMPServer.exe process is
      running on the desktop.

Client

- You must verify that the appropriate Adobe Animate
  now is installed on the client devices. The clients also
  must have IP connectivity to the media source.

Windows Client

- Install Adobe Animate now 10.1 or later for Internet
  Explorer.

Linux Client

- Install the libexpat.so.0 file, or verify that this file is already installed. Ensure that the file is installed in the /usr/lib or /usr/local/lib directory.

- Install libflashplayer.so. Ensure that the file is installed in the appropriate Flash plug-in directory for your Linux operating system.

- Install wget.

Set Up the Web Pages That Provide Multicast or Unicast Streams

- The website admin configures pages to embed multicast or unicast flash content. The JavaScript SWFObject library should be used to embed the SWF into the web page.

Prerequisites

- Verify that the swfobject.js library is imported in the MHTML web page.

- Embed the viewmp.js JavaScript command in the MHTML web page. For example:

- <script type="text/javascript" src="http://localhost:33333/viewmp.js"></script>

- Make sure to embed the viewmp.js JavaScript command before the ShockWave Flash (SWF) file is imported into the MHTML web page using the SWFObject library.

Website Flash Media Player and SWF

- The customer IT must integrate an appropriate Flash media player such as Strobe Media Playback into their website.

- To stream multicast content, you can use multicastplayer.swf or StrobeMediaPlayback.swf in web pages.

- To stream live unicast content, you must use StrobeMediaPlayback.swf.

- You can also use StrobeMediaPlayback.swf for other supported features such as RTMP streaming and HTTP dynamic streaming.

Improved HTML Access to Horizon View Desktops: Improved Access to Your Desktop from Modern Browsers

- Remote desktops delivered through HTML5-capable web browsers

- Better video playback

- More responsive typing

- Scalability with up to 350 concurrent connections through security server

Benefits

- Install-free access to desktops

- Clean, integrated, browser-based experience

- Access view desktops from device platforms where no native client is available

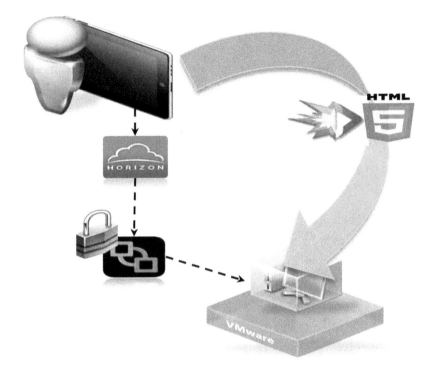

Audio Feature

- Audio playback is enabled by default. It can be disabled by using the GPO.

- Changing the overall volume level on the remote desktop does not have an impact to the volume level sent to the remote desktop. Users are expected to use the client machines volume controls.

- Audio goes out of lip-sync at times.

- With heavy traffic over limited network bandwidth or when the browser is processing I/O, it is known to have occasional choppiness/pop sound.

- The experience is much better over the Chrome browser or Firefox 25 (GA: Oct 25). For other browsers, a dialog is displayed with the following text:

"We have detected that your virtual desktop may experience audio playback issues with this browser. If audio does not play properly, please try Google Chrome."

Full-Screen/High-Res Mode

- Full-screen mode

- It's a HTML5 browser capability and works only on feature capable browsers.

- When browser switches to a full-screen mode, the browser shows a dialog requesting confirmation from the user and indicates that using the Esc key can toggle it back.

- Full-screen features do not work in IE browsers yet (no support).

- High-resolution mode.

- This feature can be used on devices that have a really high screen pixel density such that for better readability the browsers turn off the high-density pixels.

- Supported on high density pixel machines like the Retina displays on MacBook Air/Pro and Google Chrome Pixel with browser support.

- MacBook Pro 15" Retina max native resolution: 2880 x 1800

- MacBook Pro/Air 13" Retina max native resolution: 2560 x 1600

Unity Touch for iOS and Android Refinements

Even Better Integration of the Windows Desktop Experience

- Minimize apps easily

- Add favorites apps or files from search

- Single-finger horizontal scroll

- More responsive full-screen touchpad

Benefits

- Reduces frustration working with Windows on mobile devices

- Makes easier to get quick task done

- Available in iOS and Android clients

The following is a detailed breakdown of the logic/pseudocode for handling common VMware Horizon issues, including missing Nvidia drivers, and verification steps for advanced configurations.

Logic/Pseudocode for Handling Common VMware Horizon Issues

```
def check_vmware_horizon_environment():
    """
    Perform initial checks and setup for VMware Horizon
    environment.
    """
    try:
        # Step 1: Verify that VMware Horizon is installed
        if not is_horizon_installed():
```

```python
        raise EnvironmentError("VMware Horizon is not
        installed.")

    # Step 2: Check for NVIDIA GPU and drivers
    if not check_gpu():
        raise EnvironmentError("No NVIDIA GPU detected.")

    if not verify_nvidia_drivers():
        raise DriverError("NVIDIA drivers are missing or
        outdated.")

    # Step 3: Verify VM configuration
    if not check_vm_config():
        raise ConfigurationError("VM configuration is
        incorrect or incomplete.")

    # Step 4: Verify connection server and agent
    if not validate_connection_server():
        raise ConnectionError("Connection server is
        unreachable.")

    if not validate_agent():
        raise AgentError("Horizon agent is not running or
        misconfigured.")

    print("All checks passed. VMware Horizon environment is
    healthy.")

except (EnvironmentError, DriverError, ConfigurationError,
ConnectionError, AgentError) as e:
    print(f"Error: {str(e)}")
    resolve_error(e)

# Helper Functions
def is_horizon_installed():
```

```python
    """Check if VMware Horizon is installed."""
    # Pseudocode for command execution
    return run_command("Check VMware Horizon Installation")

def check_gpu():
    """Verify if a compatible NVIDIA GPU is present."""
    return run_command("lspci | grep -i nvidia")

def verify_nvidia_drivers():
    """Ensure NVIDIA drivers are properly installed and up to
    date."""
    driver_check = run_command("nvidia-smi")
    return "Driver Version" in driver_check

def check_vm_config():
    """Ensure the VM is configured with the correct
    settings."""
    # Example configuration checks
    return run_command("Verify VM Settings")

def validate_connection_server():
    """Check if the connection server is running and
    reachable."""
    return run_command("ping -c 1 connection-server")

def validate_agent():
    """Verify if the Horizon agent is running on the VM."""
    return run_command("Check Horizon Agent Service")

def resolve_error(error):
    """Attempt to resolve specific errors based on the type."""
    if isinstance(error, EnvironmentError):
        print("Please install VMware Horizon or check
        prerequisites.")
```

```python
    elif isinstance(error, DriverError):
        print("Ensure NVIDIA drivers are installed using the
        correct version.")
        print("Command: sudo apt install nvidia-driver-
        [version]")
    elif isinstance(error, ConfigurationError):
        print("Update VM settings in vSphere client as per
        VMware guidelines.")
    elif isinstance(error, ConnectionError):
        print("Check the connection server address and network
        connectivity.")
    elif isinstance(error, AgentError):
        print("Restart the Horizon agent and check its
        configuration.")
    else:
        print("Unknown error. Consult VMware documentation.")

# Verification Steps for Advanced Configurations
def verify_advanced_configuration():
    """Perform verification for advanced features like vGPU or
    Blast Extreme."""
    try:
        # Step 1: Verify vGPU configuration
        if not verify_vgpu():
            raise ConfigurationError("vGPU is not properly
            configured.")

        # Step 2: Validate Blast Extreme settings
        if not verify_blast_extreme():
            raise ConfigurationError("Blast Extreme settings
            are incorrect.")
```

```python
        # Step 3: Check SSL/TLS settings
        if not verify_ssl_tls():
            raise ConfigurationError("SSL/TLS configuration is
            insecure or incorrect.")

        print("Advanced configurations verified successfully.")

    except ConfigurationError as e:
        print(f"Advanced Configuration Error: {str(e)}")
        resolve_error(e)

def verify_vgpu():
    """Ensure vGPU profiles are configured correctly."""
    return run_command("Check vGPU Configuration")

def verify_blast_extreme():
    """Check Blast Extreme protocol settings."""
    return run_command("Verify Blast Extreme Settings")

def verify_ssl_tls():
    """Ensure SSL/TLS settings meet security requirements."""
    return run_command("Validate SSL/TLS Configuration")

# Example Command Execution Function
def run_command(command):
    """Mock function to execute system commands."""
    print(f"Executing: {command}")
    # Simulate command success
    return True

# Main Entry Point
if __name__ == "__main__":
    check_vmware_horizon_environment()
    verify_advanced_configuration()
```

This pseudocode clearly outlines the logic for handling common VMware Horizon errors and verifying advanced configurations. Key aspects include the following:

- **Error detection and resolution:**

  - Ensure prerequisites like the Horizon installation, Nvidia GPU presence, and drivers.

  - Provide actionable steps for resolving detected issues.

- **Advanced configuration verification:**

  - Includes checks for vGPU, Blast Extreme, and SSL/TLS settings

  - Raises targeted errors for specific misconfigurations

- **Reusability:**

  - Modular functions like the run_command allow customization for specific environments.

Summary

In this chapter, we went through the best-in-class products that have been integrated and optimized in VMware EUC products for the software-defined data center. There are multiple solutions to meet customer demand for on-premise, cloud, and hybrid solutions. VMware provides a complete solution including display protocol and audio and video drivers with different techniques and not point products.

CHAPTER 10

Horizon Workspace

If you take a look at the end-user computing landscape, there are apps, data, and different types of devices. Users interact with all of these. Users want to be able to get access to all their apps and data on any device of their choosing. However, the problem is that this isn't the case today. There are big distinctions between desktop and mobile, both in terms of the capabilities available to end users and how IT admins manage those devices. For end users, they want to be able to use any device to access their applications and data. They don't want any compromises in their choice of device. Admins want a simple way of managing all their employees' devices. The reality is that it takes a broad set of functionality to satisfy the demands of end users and admins. And if we take a look at the market, we can see that it's a bunch of point players (Citrix and Parallels RAS), so we will go through in this chapter how VMware can help end customers to consume IT as a service with complete end-to-end solution. We will go through different use cases in this chapter.

Solution Description: Simplify the end-user experience and reduce IT costs by combining applications and data into a single enterprise-class aggregated workspace, securely delivered on any device, with VMware Horizon Workspace.

Question: What is your security exposure if Dropbox gets hacked?

Business Issues: Enabling Mobile Workforce securely and cost-effectively

© Ajit Pratap Kundan 2025
A. P. Kundan, *Intelligent Automation with End-User Computing Solutions*,
https://doi.org/10.1007/979-8-8688-1312-2_10

Problem

- Have too many unsecured documents in services like Dropbox

- Need to securely support an increasingly mobile workforce, perhaps through a BYOD initiative

- Have too many applications and devices to support

- Befuddled by the number of point solutions, each promising to deliver part of the solution but none mastering them all

- Missing expected functionality (contributing to the perception of IT not being able to understand business requirements)

- Experiencing cost overruns (contributing to the perception that IT cannot be trusted with a budget.)

One login | One experience | Any device

Solution

- If we could provide you a single, aggregated workspace allowing your users to access their applications and data across any device and enable you to securely support a BYOD initiative, would that be of interest to you?

- If we could reduce your security exposure by providing users with a superior solution to Dropbox, would you want to learn more?

- If we could reduce your management headaches through simplified administration and enable faster provisioning, would that be of value for you?

- Would you be interested in learning more about delivering custom experiences for different users and groups all driven by policies?

- Would you be interested in working with an experienced and trusted supplier as a more strategic supplier to your business?

Value

What would it be worth to you if you could:

- Increase IT flexibility while reducing administration costs?

- Work with an experienced, proven, and trusted supplier in virtualization and ITaaS?

- Adapt and respond to changes more quickly in the business?

- Easily and securely support BYOD for the entire organization?

- Reduce exposure to data leakage?

- Reduce time to deploy new apps, functionality, or bug fixes?

Power

- CIO: Limited business impact to users, cost savings

- IT Admin: Will likely need sign off that the solution is viable

Plan

- Demonstration of the Horizon Workspace

- Build ROI and TCO model

- Case studies

- Proposal

Mobile Secure Workplace

Solution Description: The Horizon Mobile Secure Workplace is a fully validated architecture that provides end users with quick and easy access to desktops, applications, and data across devices, locations, and networks. Designed to help IT address bring-your-own-device (BYOD) and workplace mobility initiatives, this solution streamlines and automates desktop application and data management and provides IT with the ability to apply location-aware, policy-driven access to corporate resources with higher levels of security and compliance.

Question

- Are you prepared to support the multiple mobile platforms used by your employees?

- How much would a breach in security cost your company?

- Did you know that the penalty for a Payment Card Industry Data Security Standard (PCI DSS) compliance violation can be up to $200,000?

Business Issues

- Addressing BYOD initiatives

- Securing data

- Compliance with regulatory mandates such as HIPAA for healthcare and PCI for financial services

Problem

- End users use personal devices for work purposes with no IT support.

- Data tied to physical endpoints is difficult to secure and manage on mobile devices.

- Businesses must comply with regulations such as HIPAA and PCI regarding data access and storage

Solution

- The VMware Mobile Secure Workplace (MSW) solution supports all platforms and gives IT centralized management capabilities.

- MSW includes a pretested, fully validated architecture combining best-of-breed technologies from VMware with other third parties to address antivirus and security concerns.

- MSW includes solutions for compliance with regulatory mandates such as SOX, PCI, HIPAA, and FISMA.

Value

- IT is able to support end user demands for access to corporate data through the device of their choice.

- Security concerns can be met as data is moved off of endpoints and into the data center along with MSW features such as centralized antivirus and host, application, and endpoint security.

- Businesses and IT departments can rest assured that they are meeting regulatory standards without fear of being fined.

Power

- VP of IT: Responsible for the management of IT infrastructure and operations

- CIO: Responsible for managing budgets, maintaining business alignment, implementing security, being compliant, managing resources, managing customers, and managing change.

- Director of End-User Computing: Tasked with the delivery of applications and services and the integration of end users into the computing environment

- Lead Architect: Responsible for planning, designing, implementing, and operating enterprise data center infrastructures, and often includes virtualization.

Plan

- Discover size and distribution of user population, devices, and physical locations

- Identify compliance requirements

- VMware PSO to assist with architecture and implementation of solution

Image Management

Solution Description: You need a way to more effectively manage and deliver Windows and applications.

Question:
How much IT time is spent on reactive operations instead of proactive operations, and is that affecting the quality of service you are able to deliver to your end users?

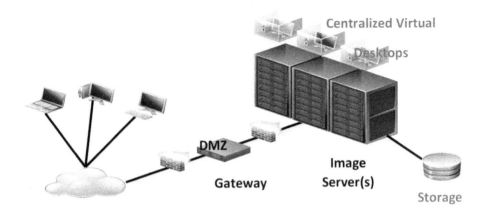

Business Issues

These are the typical issues impacted by this solution that the buying audience needs to address and resolve to contribute to the overall company objectives. They normally have revenue or profit impact. They are typically a metric that the audience will be measured on. Three common issues are time to market, cost management, and quality/reliability.

Problem

- Are you facing a lack of resources to perform critical IT operations at an acceptable level of satisfaction to your end users?

- Do you find yourself managing a plethora of Windows images and applications without any easy way to update them and deploy them?

- Is it a challenge for your end users when you perform updates to their Windows OS or applications that result in them losing data, user-installed apps, or profile customizations?

- Do you have an easy way to roll back a specific part of a malfunctioning end user's PC to fix the problem without affecting other pieces?

Solution

- If I could show you a solution that categorized end-user PCs into logical layers that allowed IT to manage, deploy, and fix only the parts IT wanted to manage, without affecting any of the other user data, would this be of interest to you?

- If we could also provide a solution that allowed you to do centralized break-fix on any end user's PC directly on the logical layer that was malfunctioning, would this decrease your help-desk cost?

- Do you have branch office locations that would benefit from a simpler way to deploy Windows images and updates, and applications or packages, over a WAN optimized protocol without requiring any branch infrastructure?

Value

- The VMware solution uses a patented centralized image approach that provides labor cost savings from 45% to 85% (figures based on Gartner research).

- The VMware solution will significantly reduce deployment time for a new user or a user that is in need of desktop recovery.

- With VMware, IT personnel do not need to set up infrastructure at each branch or physically assist with IT operations at each location. All work is managed centrally from one console.

- Using VMware, risk is effectively mitigated by leveraging a rolling snapshot that allows IT to rollback users to a previous working state.

- End-user downtime and help-desk support time are greatly reduced.

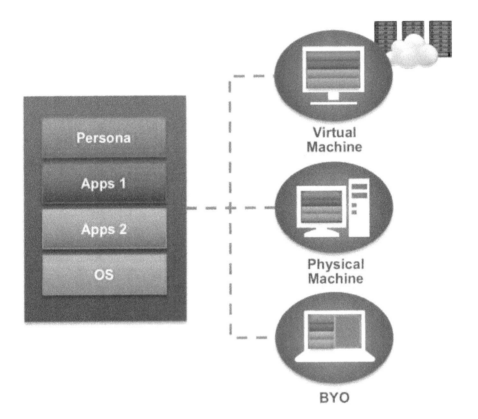

Power

- CIO: Limited business impact to users, cost savings

- IT admin: Will likely need sign off that solution is viable

Plan

- Discover size and distribution of user population, devices, physical locations

- PSO to assist with architecture and implementation of solution

- Monthly status checkpoint meeting until solution is deployed

Windows 11 Migration

Solution Description: Need to deliver effective cost management, while maintaining high level of business agility during a **Windows 11 migration.**

Question: What are the consequences if you and your team do not get all users migrated to Windows 11 in a timely, cost-efficient manner while keeping users productive?

Business Issues: These are the typical issues impacted by this solution that the buying audience needs to address and resolve to contribute to the overall company objectives. They normally have revenue or profit impact. They are typically a metric that the audience will be measured on. Three common issues are time to market, cost management, and quality/reliability.

- **Problem:** Are you facing a lack of resources to complete the migration from Windows 7 to Windows 10/11 before Microsoft discontinues support?

- Will your business tolerate multiple hours or even days of productivity outage or degradation due to the migration?

- Is your user population highly distributed geographically, complicating the logistics of a migration?

- Do you have a strong risk mitigation plan if a user's migration fails?

Solution

- If I could show you a solution that would provide you an automated migration for laptops and desktops in any location, would that be of interest to you?

- If we could also provide this zero-touch solution such that it would result in less than one hour of user downtime and maintain all user personalizations, would that be compelling for you?

- Do you have multiple branch locations that would benefit from an approach that is especially strong in managing the remote users/devices that are geographically dispersed?

- Would it be of value if you can roll a user back immediately to a previous state if anything should go awry with the migration?

Value

- The VMware solution uses a patented centralized image approach that provides labor cost savings from 45% to 85% (figures based on Gartner research).

- The VMware solution will also significantly reduce deployment time from an average of 2.5 to 4 hours to just 30 to 45 minutes per user while preserving user personalizations, allowing users to maintain a high level of productivity.

- With VMware, IT personnel do not need to set up infrastructure at each branch or physically assist with migrations at each location. All the work is managed centrally, in a zero-touch fashion.

- Using VMware, risk is effectively mitigated in situations where migration fails, maximizing user productivity during a disruption like OS migration.

Power

- VP Operations: Needs to provide a smooth migration with least cost possible, while providing little interruption to users

- CIO: Limited business impact to users, cost savings

Plan

- Discover size and distribution of user population, devices, physical locations

- Develop migration plan given user population

- VMware PSO to assist with architecture and implementation of solution

- Monthly status checkpoint meeting until migration is complete

Branch Office Desktop

Solution Description: The Horizon Branch Office Desktop provides a comprehensive approach to addressing multiple requirements within the branch office. Leveraging this solution, IT organizations can centrally manage OS images for both their physical end points and their virtual desktop environments, while ensuring employees have fast, secure access to the applications and data they need to maximize productivity.

Question:

- How much does it cost to manage each of your remote locations?

- What would be the consequences if there was a breach of security at one your remote locations?

- How would your business be impacted if a remote site had an outage or suffered from poor performance over the WAN?

Business Issues

- High cost of managing remote office or branch office locations

- Securing data

- High availability and WAN performance

Problem

- High cost of managing and supporting remote locations with on-site visits.

- Data tied to endpoints for industries with sensitive IP makes security and compliance a big challenge, as it can easily go missing.

- Poor WAN links and unreliable connectivity lead to poor desktop and application response times and inconsistent desktop experience for end users

Solution

- The VMware Branch Office Desktop (BOD) solution is a fully validated architecture that helps IT centrally manage desktops and images for branch-office employees. With IT centralized, this solution ensures that IT can streamline management and reduce the costs of supporting end users across distributed sites.

- This solution centralizes all data to better protect corporate data.

- BOD ensures that end users experience a reliable desktop experience across the WAN by leveraging VDI appliances in the branch and Horizon technology.

Value

- Centralized management of virtual desktop images and desktops reduces opex and IT complexity and ensures end users can get back up and running quickly in the event of an outage.

- Security and compliance mandates can be met as data is moved off of endpoints and into the data center.

- Users get the best possible experience with LAN-like performance across the WAN for fast application response times so they can get on with their work in a predictable and efficient manner.

Power

- VP of IT: Responsible for the management of IT infrastructure and operations

- CIO: Responsible for managing budgets, maintaining business alignment, implementing security, being compliant, managing resources, managing customers, and managing change.

- Director of end-user computing: Tasked with delivery of applications and services and integration of end users into the computing environment.

- Lead Architect: Responsible for planning, designing, implementing, and operating enterprise data center infrastructures, and often includes virtualization.

Plan

- Discover size and distribution of users and physical locations

- Identify compliance requirements

- VMware PSO to assist with architecture and implementation of solution

Business Process Desktop

Solution Description: Deliver desktops as a managed service to contact center agents, offshore developers, and outsourced workers. The VMware Horizon View Business Process Desktop solution drives down operational costs and provides end users with a highly responsive secure computing experience across networks and locations.

Question:

- How would your customers react if they found out that a third party had access to their data?

- How would your business be impacted if your contact center operations became inaccessible?

- How can you support upgrades and provisioning of new users to offshore locations?

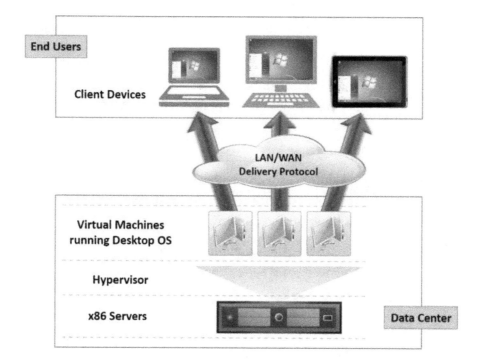

Business Issues

- Secure access to data and applications by third parties or offshore locations

- High availability of offshore operations

- Ease of management of offshore or third-party locations

Problem

- Offshore and outsourced workers may compromise corporate security.

- Many offshore locations lack modern IT infrastructures and are more susceptible to outages.

- Provisioning new sites, new users, and contractors takes too long, and troubleshooting remote users is difficult and costly.

Solution

- The VMware Business Process Desktop (BPD) solution includes a pretested, fully validated architecture combining best-of-breed technologies from VMware with other third parties to address antivirus and security concerns.

- The BPD solution provides end users with a highly available architecture that delivers virtual desktops across multiple trusted and untrusted locations.

- BPD centralizes all data enabling fast provisioning, disaster recovery, and easier upgrades.

Value

- Reduce risk and have greater control of data security and compliance for third parties.

- Ensure that offshore and outsourced operations have access to data and applications at all times.

- Reduce the time needed for deployment of new locations and provisioning of end users from weeks to minutes.

Power

- VP of IT: Responsible for the management of IT infrastructure and operations

- CIO: Responsible for managing budgets, maintaining business alignment, implementing security, being compliant, managing resources, managing customers, and managing change.

- Director of End-User Computing: Tasked with delivery of applications and services and integration of end users into the computing environment.

- Lead Architect: Responsible for planning, designing, implementing and operating enterprise data center infrastructures, and often includes virtualization.

Plan

- Discover size and distribution of users and physical locations

- Identify compliance requirements

- VMware PSO to assist with architecture and implementation of solution

AlwaysOn Desktop

Solution Description: Industries such as healthcare, government, first responders, and financial services demand immediate and constant access to desktops and applications to do their jobs. The VMware AlwaysOn Desktop is a solution architecture that provides continuous availability of virtual desktops and applications across devices, locations, and networks. This solution is an active-active configuration with multiple levels of redundancy as well as continuous monitoring and load balancing between sites.

Question:

- Can you provide constant availability to your desktops, applications, and data...even in a disaster?

- Are you able to support access to data and applications from mobile devices?

- How much would a breach in security and compliance cost your company?

Business Issues

- High availability and disaster recovery

- Mobility and session management

- Security and compliance

Problem

- Desktop high availability solutions are complex and hard to implement.

- Mobile access to data and applications is required for first responders and healthcare professionals.

- Industries such as healthcare, government, and financial services require compliance with regulatory mandates such as HIPAA, FISMA, and PCI, respectively.

Solution

- The VMware AOD (AlwaysOn Desktop) provides end users with constant availability to applications and data through a centrally managed private cloud-based desktop solution (no data on device).

- The AOD solution provides a consistent end-user experience with high-performance access to all apps and data from any device.

- AOD includes solutions for compliance with regulatory mandates such as SOX, PCI, HIPAA, and FISMA.

Value

- AOD provides 24/7 uptime, reduces risk, and gives IT greater control of data security

- IT is able to support end-user demands for access to apps and data through the device of their choice.

- IT departments can rest assured that they are meeting regulatory standards without fear of being fined.

Power

- VP of IT: Responsible for the management of IT infrastructure and operations

- CIO: Responsible for managing budgets, maintaining business alignment, implementing security, being compliant, managing resources, managing customers, and managing change.

- Director of End-User Computing: Tasked with delivery of applications and services, and integration of end users into the computing environment.

- Lead Architect: Responsible for planning, designing, implementing and operating enterprise data center infrastructures, and often includes virtualization.

Plan

- Discover size and distribution of users and physical locations

- Identify compliance requirements

- VMware PSO to assist with architecture and implementation of solution

Hardware Vendor and VMware: PC Refresh

Solution Description: A joint hardware vendor and VMware desktop virtualization solution that allows IT organizations to extend the life cycle of existing hardware while moving to a platform that allows for centralized and secure management of desktop assets in the data center. End users can also enjoy a better desktop experience across existing/older hardware while seamlessly accessing their applications and data across devices and locations.

Question:

- Are you finding that refreshing your fleet of PCs is continually consuming the bulk of your budget?

- It can take end users roughly 15 minutes to log onto a five-year old device. That's more than an hour a week of lost productivity across each employee. Is downtime like this something that your department can cost justify around your older devices?

Business Issues

PCs need to be refreshed every three to four years in order to perform effectively. This ends up costing organizations a lot of money and as the machines age, which means that user productivity is impacted, which can affect a company's top line.

On average it takes end users about 15 minutes to log into a 5-year-old device, which amounts to roughly 1 hour per employee per week in lost productivity.

Problem

- PCs are costly to refresh every three to four years.

- Most customers tend to want to extend this, but as they do, user performance on these devices is impacted.

- Older devices mean slower log on times and also end up taking more resources and time to manage and maintain. With older machines, there are a lot more break-fix issues that IT needs to deal with.

- Physical machines are also noisy and consume a significant amount of power (9X that of thin or zero clients).

- With these physical devices, IT can spend up to 8 hours deploying a single desktop and over 4.5 hours a year patching and updating user applications across each device.

Solution

- VMware Horizon View with Hardware Vendor servers and storage provide an end-to-end desktop virtualization platform that allows customers to centrally and securely manage desktops and assets in the data center while allowing end users the freedom to access their customized virtual desktops across devices and locations.

- Coupling this solution with the hardware vendor Wyse thin clients allows customers to save on power, simplify desktop management, and reduce endpoint costs as thin clients typically last three to four years longer than PCs.

Value

- IT departments can get started with VDI very cost-effectively by extending the lifecycle of the devices they have on hand and taking the money they would have spent on new PCs and putting it into the virtual desktop infrastructure needed to support end users.

- Moving to VDI allows customers to centrally and securely manage images and desktops, reducing opex by as much as 50%.

- Replacing very old devices over time with Wyse PCs allows customers to further reduce power costs, simplify management, and reduce endpoint costs as thin clients last for three to four years longer than physical PCs.

Power

Head of IT

Plan

Phase-wise deployment with defined dos and don'ts

Modernize the Desktop for Better Patient Care

Healthcare Technology Disruption: We are watching what will be the fastest transformation of a major U.S. industry in history. We can also use virtualization to help solve the desktop management and availability issues.

We can use View to separate the OS from the underlying physical devices, and we can use Thin App to separate the apps from the OS and keep the user data and settings persistent. Then we move them into the cloud or data center where they can be managed independently of each other. We can patch the OS without worrying about the apps, and we can update the apps without worrying about the OS all while maintaining user data and settings.

When a user wants to access their point of care desktop, we can dynamically assemble these components back together to present them with a single and unified view of their point of care desktops with all of their applications and data readily available to them.

Since the desktop is being delivered as a managed service, we can use any qualified device to access it.

VMware Vision for Healthcare: Nonstop Point of Care

- Clinical devices are Tier 1, meaning access, security, availability, and mobility.

- Enable IT to deliver clinical desktop computing as a reliable managed service.

- Provide a better experience for the caregiver.

- Enable caregiver mobility and secure access to patient data.

- Cloud-based delivery platform for desktops and clinical applications.

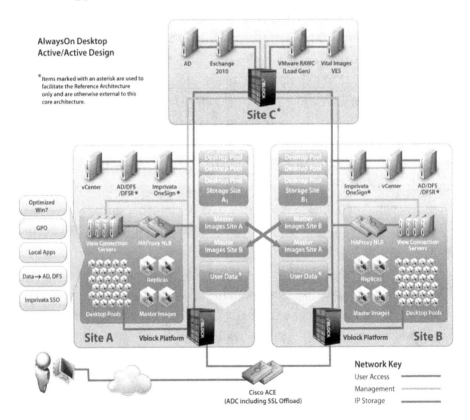

AlwaysOn Point of Care: reference design

- Practice management/EHR services

- Virtualized desktops, office productivity, mail

- Imaging, dictation, billing

Empower Caregivers with always-on, anywhere access

Solution Elements: Validate Reference Design

AlwaysOn Point of Care is a validated solution architecture from VMware. It was specifically built to meet the needs of healthcare organizations. It combines VMware and ecosystem products and services to meet the stringent requirements for availability, security, elastic scaling, rapid and automated provisioning, high capacity, and low latency with no single point of failure. Key solution elements include VMware View, which is the cornerstone of this solution.

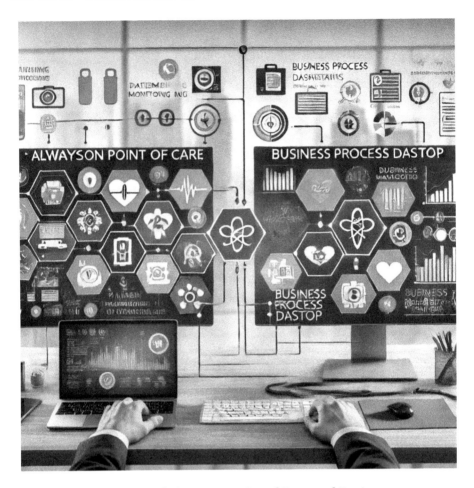

Interconnected systems of AlwaysOn Point of Care and Business Process Desktop

Preconfigured infrastructure platforms are sized to support specific workloads. These platforms are available to run both on the customer site and in a hosted (cloud) facility. The integrated stack reduces integration and testing time to speed deployment efforts. Scalability and performance are designed into each platform, so adding more desktops is simplified. The Unified Infrastructure Manager provisions services on the platform in minutes, with a few mouse clicks, deploying VMware clusters quickly and consistently. These platforms are available in a variety of models to fit any VMware View requirements, from hundreds to thousands of desktops.

Load balancing: The AlwaysOn Point of Care solution also provides a load-balancing application delivery controller. The user connects to a virtual IP address configured on the device. The health monitoring of VMware View Connection Servers provides the user with high performance and availability. This helps offload the CPU-intensive tasks from the broker. With the High-Availability (HA) feature enabled, the connection replication happens by default

Authentication: SSO and strong authentication permit users to access all workstations and applications they are authorized to use. By configuring and linking multiple instances of virtual appliances at both sites with fault tolerance, during the site failover, desktop agents can continuously look up the next working instance without disrupting the workflow.

Storage: Storage offers reliable high performance. Site-to-site replication of the master images is performed with Replicator to provide data redundancy and reduced recovery times. Asynchronous replication is used with NFS datastores. The storage platforms have the flexibility to support a variety of replication models, including synchronous or asynchronous block-based replication.

Summary

The primary goal of every person in IT is to offer the best experience to all of their peers. Enable them to carry out their everyday actions without worrying about the nuances. That has become especially challenging with the ever-evolving app ecosystem. You can host a website where users can download a few of the apps needed, and they can walk to IT to get access to some licensed apps, use remote app information from the website, use browser bookmarks, and get a few from an app store. In other words, as an end user, you have to be aware of the type of application needed, remember the source, and go fetch it. Oftentimes you have to search all the sources to find what you need. Now imagine you have one button on a dashboard/desktop that contains everything needed. That is what Workspace ONE offers. It also has insight into the actual progress of the application, providing the ability to display the progress to your end users and making the solution ideal for self-service situations as well. In this chapter, we went through all the relevant functionality/use cases including healthcare.

CHAPTER 11

Latest Capabilities for Android Deployments with VMware

In this chapter, we will discuss how Android devices can work with VMware end-user computing solutions and all the current updates/ features that you can leverage in Android-based deployments. We will also discuss how Citrix XenApp can leverage Horizon to provide a unified console for end users so they can access their all required/authorized applications easily through the native GUI.

It's no great mystery that Android is absolutely dominating the mobile OS market, and not just by a small margin. In fact, Android is almost five times as prominent as all other mobile platforms combined. For years we've seen Android become ubiquitous across all embedded platforms, and yet I'm still having conversations with customers on a regular basis who are hesitant to embrace Android to the extent that they've embraced today's iOS and yesterday's BlackBerry. If you haven't gone all-in and adopted Android as your primary mobile strategy (and I'm about to tell you why you should), you at least need to have a solid support plan for it.

© Ajit Pratap Kundan 2025
A. P. Kundan, *Intelligent Automation with End-User Computing Solutions*,
https://doi.org/10.1007/979-8-8688-1312-2_11

And when I say that Android has become ubiquitous, I truly mean it. My good friend Andrew Toy from Google uses the term "glowing rectangles" to refer to Google's goal of having Android be the de facto platform for any device with a screen and a microprocessor. I think they've all but reached this goal.

Common Misconceptions Surrounding Android in the Enterprise

- Fragmentation across features

- Difficult provisioning

- Lack of end-user privacy

- Vulnerable security

When I talk to those IT professionals and security folks who are still reluctant to bet big on Android, I consistently hear four common themes:

- I hear the dreaded term "fragmentation," meaning IT has the perception that features are so inconsistent across various Android OEMs and devices that trying to build consistent policies around them is too burdensome.

- I hear that the barrier to entry for end users is too high, and IT struggles to aid users in self-service provisioning.

- I hear that the market is still struggling to gain widespread adoption of BYOD and COPE programs and that their end users are still concerned over their personal privacy

- Finally, I hear the misconception that Android devices are simply insecure and open to too many vulnerabilities.

I've done so much work alongside Google over the last couple of years so I know that none of these should be excuses any longer.

Closing the Fragmentation Gap

Let's start by talking about fragmentation.

What Has "Fragmentation" Traditionally Meant?

- Inconsistent support and capabilities per OEM

- Varying ability to configure and manage mail clients

OEM and Carrier OS Customizations

Lack of standard software update/patch cycles: Up until 2015, the term "fragmentation" was synonymous with Android, and it became the thorn in the side of most IT organizations driving mobility programs. It became an "excuse" to put Android on the back burner and focus on other platforms, much to the chagrin of a plurality of end users (those hard-core Android believers) and much to the detriment of the future mobile estate within these organizations.

Still others approached fragmentation head-on and solved it by standardizing on a single OEM or model. They may have fared pretty well for their corporate-owned deployments, sometimes at significant financial cost, but struggled in the BYOD arena. Try convincing your Nexus-toting engineer to switch his personal phone to a phablet.

Google for Work | Android

Google for Work integrates with Android for Work for a consistent experience across manufacturers to provide personal and work profiles in a single, unified launcher. Beginning in 2015, vanilla Android provides a standardized management platform focused on separation of work and personal data and allows organizations to truly embrace the power of Android without worrying about which devices support certain features; they all do. For BYOD and COPE deployments, the Work Profile lets the end user maintain complete control over their device and visually separates their work apps from their personal apps, even allowing them to run the same apps in both camps. IT's control is limited to only the business apps and data, and all the way down to the hardware level, the business and personal profiles are separated from each other. Android for Work is the only truly comprehensive OS-provided containerized management solution in the ecosystem today.

And with Android for Work's Device Owner mode, corporate-dedicated and line-of-business deployments are a breeze. A customized launcher and curated app list allow the device to be as locked down or as open as IT chooses to permit. It is all built in to the OS.

I even worked with Google to standardize on a manageable native email client within Android for Work. AirWatch now supports the ability to configure and manage your Exchange accounts directly in the Gmail app. And nobody in the EMM space has a better relationship with Google and the Android for Work team. As one of the original launch partners for Android for Work, VMware engineers continue to work with the Android team on a daily basis to continue to perfect the Android for Work platform.

What Does "Fragmentation" Mean Now?

So now that we've put the fragmentation battle to bed, I want to introduce the next wave of fragmentation, and ironically enough, it's Android that is the only platform ready to tackle it.

With the continued growth of the Internet of Things, Android has proven itself as the only standard cross-form-factor operating system. From your mobile phone to your vehicle, you're probably touching more than a few Android-powered devices every day. And every one of them has the opportunity to be managed by AirWatch. So the next time you embark on that next big mobile IT program, think about how you can avoid fragmentation by standardizing on Android.

It's not about fragmentation across manufacturers anymore. Now it's about fragmentation of form factors.

- Handsets

- Tablets

- Wearables

- Augmented reality/glasses

- Connected cars

- GPS receivers

- TVs

- Appliances

- And so many more...

Android is the only platform by which all form factors are powered.

Simplifying Registration and Onboarding

Now, for those of you who believe that getting up and running with Android is still a struggle, I'm excited to show you what we've been working on over the last year to bring Android deployments down to just a few minutes. Since the initial launch of Android for Work in 2015, we've worked with Google and customers like yourselves to streamline the setup process. Two primary areas of improvement that we identified right off the bat were the admin setup process and the end-user onboarding process. The ecosystem was very clear that Android for Work's enterprise domain claiming and requirements for managed Google accounts were far too complex just to get the first device enrolled. We listened. We collaborated with Google and built an entirely new Android for Work process around Google's concept of Android for Work accounts. Now administrators only need a simple Gmail account and 30 seconds to get their organization up and running. And end users no longer need to worry about being prompted for a Google account, which may seem foreign to anyone who doesn't use Google Apps for Business.

Let's take a look at a brief demonstration of the new admin setup and enrollment processes. Over the next couple of minutes, Eric is going to register his enterprise with Android for Work and then enroll his first device. Now, Eric is the IT administrator for ACME Enterprises, and he wants to get his BYOD Android devices enabled and secured for access to corporate resources. He first needs to establish his enterprise with Google, and all he needs is a simple Gmail account. From the AirWatch console he'll click the configure button. He's quickly redirected to a Google registration page where he'll sign in with his Gmail account and provide his company's name. After confirming, he's redirected back to the AirWatch console, and his registration is complete. He's now built a private management channel in the Google cloud.

Try getting up and running this quickly with any other device platform.

Simplified Distribution of Apps with Google Play for Work Integration

- All the power of Play for Work, embedded in console
- New UI
- Faster search

In addition to the onboarding challenges we heard from the field, there was some resounding feedback that managing enterprise apps in Google Play for Work was too challenging. Administrators were managing their devices and policies in the AirWatch console, searching for and assigning their enterprise apps in the Google Play for Work console, and then pulling it all back together in AirWatch. We listened. We worked with Google in an effort to collapse all of the goodness and power of Play for Work to bring it right into the AirWatch console.

Let's look at an example of someone leveraging the power of AirWatch's embedded Google Play for Work controls.

Say back at ACME Enterprises, Eric wants to make sure that all of his sales reps have access to the Salesforce1 app. Instead of opening a separate console for Google Play for Work, Eric will simply add a new public app from the AirWatch console, and you'll see how we've embedded the Play for Work search and approval process right there in the console. As soon as he's done finding Salesforce1 and approving it in his enterprise, he'll leverage AirWatch to assign the app to his smart groups and define all other deployment settings. AirWatch is communicating with Google Play in the background to assign the app to the proper users and devices.

When Eric picks up his device, he can open the Play for Work store and see the app available to him.

Bulk Enroll Knox Devices with Minimal Interaction

- Turn on the device.

- Set up Wi-Fi .

- Finish the Setup Wizard.

- Deal with the MEP pop-up.

- Following the EULA and prompts.

- The device is enrolled.

That's what we've been working on with Google. We've also been working very closely with Samsung to help streamline the enrollment process there.

One of the most important facets of corporate-owned device programs is the requirement that devices should always be enrolled, even if the devices is factory reset. Moreover, IT doesn't want to have to provide white-glove assistance to help end users enroll. So we worked with Samsung to integrate KNOX Mobile Enrollment. KNOX Mobile Enrollment gives AirWatch the ability to enroll the device right from the out-of-box Setup Wizard. The end user is walked through the enrollment process as they're setting up their device for the first time, providing a fool-proof way to get from factory state to enrolled in just a couple of minutes.

As soon as Eric takes his device out of the box, he'll first set up his Wi-Fi configuration. Once he finishes the Setup Wizard, he'll see a Mobile Enrollment prompt, then the familiar AirWatch agent prompts, and then enrollment completes. Eric didn't have to follow any prescribed documentation or processes. The Samsung device itself holds his hand through the entire process.

Seamlessly Connected EMM in the Knox Workspace with VMware Tunnel

Beyond the native security capabilities that Android provides, we've also worked strategically alongside Samsung to provide extended capabilities within the KNOX Workspace. We hope all of you are familiar with the KNOX Workspace, and if not, I encourage you to visit our Samsung colleagues outside in the MOBILITY EXPO area to learn more. KNOX Workspace is designed around high-security deployments and is yet another few layers of security above and beyond the native Android feature set. In the spirit of high security and containerization, we've collaborated with Samsung to enable our AirWatch Tunnel per-app VPN functionality from within the KNOX Workspace. We aim to make it incredibly simple to leverage Samsung's on-device security and pair it with AirWatch's secure tunneling, right out of the box. We believe this is the penultimate solution for those ultra-high security deployments.

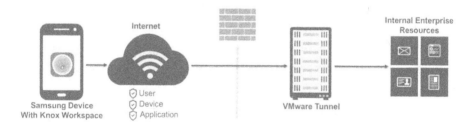

Protecting data with multilayered security

Enabling BYOD and End-User Privacy

Now at the beginning of this presentation, I showed you a chart depicting the pervasiveness of Android devices globally. Being that Android represents such a massive majority of consumers' devices across the world, it's now truly a requirement to support these devices in BYOD programs. The challenge here is that end users want to be reassured that their personal data is kept private, so it's important to implement a solution that inherently builds separation between work and personal data.

Enable BYOD with Workspace ONE

- Integration with Android for Work Profile Owner Mode
- Adaptive management for sensitive company data
- Access to work-critical web, native, and internal apps

VMware Workspace ONE for Android is designed exactly on these principles. I'm sure that you have heard a lot about Workspace ONE for the last year. The idea behind Workspace ONE is that the device is not managed whatsoever but rather presents the end user with a list of available apps and grants access to each app based on the management posture of the device. For example, if the user simply wants to reach a browser-based application, then they can do so easily without any management at all on their device. However, if they want to install a native Android app that includes sensitive company data or access email natively on their device, they'll be required to step up their management and security posture before they can install and run the app.

Deploy Android for Work Profile Owner

- Separate work and personal apps and data
- Wipe only data in the Work Profile
- Enable end-user privacy

Stepping up the management posture within Workspace ONE simply means enabling Android for Work's Profile Owner. As I mentioned earlier, with Android for Work Profile Owner, IT maintains no control over the device itself but rather over a managed container of enterprise apps. Within this managed profile, the user can install enterprise apps and email and rest assured that their personal data is inaccessible by IT. However, IT reserves the right to remove access to this Work Profile if the user falls out of compliance or leaves the organization.

And the best part about Android for Work Profile Owner is that the user never sees any scary prompts alerting them that the company can wipe their device or enforce heavy policies and restrictions. With Profile Owner, the end user intuitively understands that their privacy is being respected and appreciates the ability to access corporate resources with the best possible user experience.

VMware Workspace ONE and Android for Work provide the most optimal solution for BYOD today.

Provide Security and Productivity with VMware Boxer

For End Users

- Integrated productivity workflows

- Combined work and personal views

- Support for out of office, editing calendar responses, and reply/forward calendar events

For IT

- Secure email container with or without any device-level management

- Containerization strategy

- DLP controls and full encryption

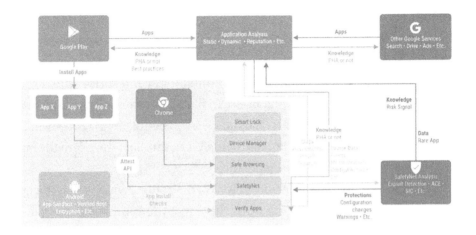

Finally, when it comes to BYOD, we know there are many cases where end users simply want work email on their personal devices. If mail is the only thing that the end user wants, then there's no need to complicate the user experience. To provide a secure, managed, email-only experience to end users on Android, VMware Boxer is the best possible way to do that.

For IT, Boxer offers a best-in-class email container complete with encryption and DLP. For end users, Boxer was developed with the end-user experience in mind first and has for years been considered one of the top consumer email clients in the market.

Protecting Data with Multilayered Security

Finally, I will spend just a bit of time touching on the security aspects of Android. In the early days of Android, the perception was that because Android was such an open platform, it was ridden with malware and exploits and a real threat to enterprises. The truth of the matter is, Google has been driving best-in-class security not only within the OS itself but throughout the Android ecosystem of OEMs and app developers. And AirWatch has been working closely with Google to implement all of the EMM-managed controls available.

A Common Framework for Any Screen and Any Use Case

Strengthened Security: Security is at the core of all we do.

Improved management capabilities: VMware engineers work hard so IT admins don't have to.

Increased control for employees: We want happy and productive employees.

Security is the core of all we do.

Industry leading protection: on device protection and Google services

Trust and Security Built In to the Device

- Device encryption

- Fingerprint recognition

- Permissions

- Safety browsing

- Verified boot

Direct Boot and Encryption

New file-based encryption allows the device to boot and run some services without the user data being decrypted.

Work security challenge

IT admins can now set password policies on work apps so employees won't have to enter complex passwords each time they unlock their device.

Always-on VPN

The IT admin can lock down network activity from boot to shutdown to ensure that work data is secure while in transit.

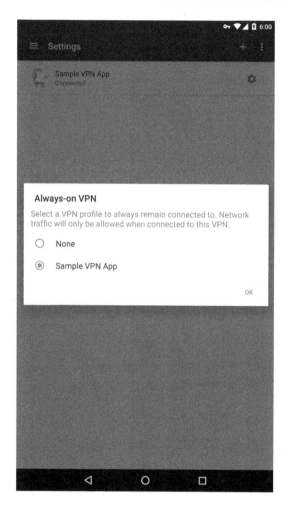

Work profiles for BYOD, CYOD, and COPE

- Dedicated work profile isolates work data in "work" folder

- Seamless transition between badged work apps and personal apps

- EMM/IT only manage work data; can't erase or view personal content

Management Highlights Single-Use Mode for COSU

- Locks device down into special mode running single or set of specific apps

- Admins can remotely control settings and app install/uninstall over the air (OTA)

- Provisioning device as simple as a near field communication (NFC) bump

Fast and Secure Setup

- With Setup Wizard for work now supporting QR code scanning, setting up Work Managed Devices can be significantly faster.

Google Play for Work

- Allows IT to securely deploy and manage business apps

- Any app in the Play catalog to be deployed to the work profile

- Simplifies app distribution and ensures IT approves every app deployed

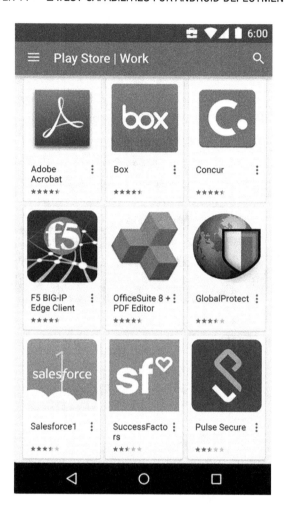

Management Highlights: Policy Transparency

- IT admins can provide full disclosure for all IT-enforced policies and restrictions so every employee understands precautions taken to keep the device secure.

Increased Control for Employees

VMware engineers and Service providers want happy and productive employees.

> **Work mode**: By turning off Work mode, employees can completely disable their device's Work Profile, including apps, notifications, and background sync so they can set their own work-life boundaries.

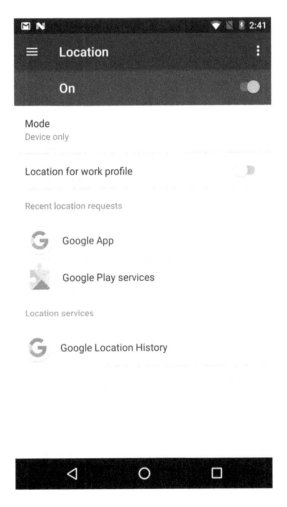

Location off switch: Employees can disable location services for apps in the work profile.

AI plays a significant role in enhancing Android management with VMware, particularly through the VMware Workspace ONE platform, which integrates AI-driven capabilities to streamline and optimize device management. Here's how AI contributes to Android management in this context:

1. **Automated Device Enrollment and Provisioning**

   - **AI-powered workflows** simplify the enrollment of Android devices by automatically applying predefined policies and configurations.

   - **Predictive algorithms** ensure that devices are provisioned with the correct profiles, applications, and settings, minimizing manual intervention.

2. **Improved Security and Compliance**

   - **Anomaly detection:** AI monitors device behavior to identify unusual activity or potential threats, such as unauthorized access or malware.

   - **Predictive compliance enforcement:** AI ensures devices remain compliant with security policies by proactively detecting and addressing potential violations before they escalate.

3. **Enhanced User Experience**

   - **Personalized recommendations:** AI analyzes user behavior to suggest frequently used applications or settings adjustments.

- **Self-healing devices:** AI detects issues (e.g., app crashes or connectivity problems) and automatically implements fixes or provides actionable guidance to users.

4. **Efficient Resource Management**

- **Predictive analytics for app performance:** AI tracks app usage and resource consumption to optimize performance and recommend app updates or removals.

- **Battery optimization:** AI monitors battery usage patterns and adjusts settings or notifies users of potential drains.

5. **Advanced Insights and Reporting**

- AI processes large volumes of data to generate actionable insights, helping IT administrators make informed decisions about Android device management.

- Real-time dashboards and reports powered by AI highlight key trends, such as app adoption rates or security risks.

6. **Proactive Troubleshooting**

- AI can identify and resolve issues before users report them. For example, Workspace ONE's AI can flag connectivity issues or configuration errors and initiate remediation steps automatically.

- Predictive maintenance alerts IT teams about hardware or software problems that might arise, allowing for preemptive action.

7. **Zero Trust Security Implementation**

- AI helps implement and enforce a zero-trust security model by continuously verifying user identities and device states.

- Risk-based authentication decisions are automated, ensuring secure access to corporate resources.

8. **Scalability and Automation**

- AI enables the management of large fleets of Android devices by automating repetitive tasks like patch management, app deployment, and update scheduling.

- Intelligent automation ensures consistent device performance and security, regardless of scale.

By leveraging AI, VMware's Workspace ONE platform delivers a smarter, more efficient approach to managing Android devices, reducing administrative overhead while enhancing security, user experience, and operational efficiency.

XenApp to Horizon: Four Simple Options to Move Forward

Integrate XenApp with Horizon

- One web portal for Horizon virtual desktops, SaaS applications, and XenApps

- Replaces Citrix Web Interface or StoreFront with VMware Identity Manager web portal

- Maintains existing XenApp RDS hosts and Citrix Receiver client

Migrate XenApp to Horizon

- Converts healthy XenApp RDS hosts with installed applications to Horizon

- Replaces XenApp Web Interface or StoreFront with VMware Identity Manager web portal

- Replaces Citrix Receiver with Horizon Client

Replace XenApp with Horizon

- Implements new Horizon RDS Hosts with installed applications

- Replaces XenApp Web Interface or StoreFront with VMware Identity Manager Web portal

- Replaces Citrix Receiver with Horizon Client

Enhance XenApp with Horizon and vSphere

- App volumes: Real-time application delivery with lifecycle management

- User environment manager (UEM)

- ThinApp application virtualization

- Horizon virtual desktops

- vRealize Operations for published apps

- vSphere, Virtual SAN, NSX

Integrate XenApp with VMware Identity Manager

- Deploy VMware Identity Manager.

- Implement the Integration Broker.

- Configure the Identity Manager for XenApp.

- Decommission the Citrix web interface or storefront.

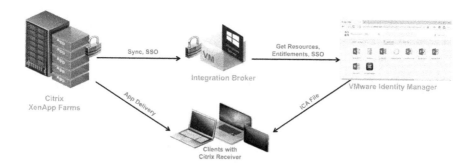

Migrate or Replace Citrix XenApp with VMware Horizon

- Deploy VMware Horizon.

- Deploy VMware Identity Manager (optional).

- Prepare Remote Desktop Services hosts.

- Create RDSH farms and application pools. Entitle users.

- Decommission XenApp.

XenDesktop: Migrate to the Next Generation of Virtual Desktops

Migrate XenDesktop to VMware Horizon

- App volumes: Real-time, personalized application delivery

349

- User environment manager

- Replace Citrix delivery controllers with Horizon

- Replace Citrix EdgeSight with vROps for Horizon

- Decommission PVS/MCS

- Migrate web interface and storefront to VMware
  Identity Manager

Enhance XenDesktop with Horizon and vSphere

- Real-time, personalized application delivery

- User environment manager

- Application virtualization

- vRealize Operations for published apps

- vSphere, instant clones, virtual SAN, and NSX

- PVS/MCS: optional or replace with Instant Clones

- Dedicated desktops ➤ Disposable

- Persistent desktops ➤ Generic

- Numerous desktop images ➤ Single

- Personalize generic desktops in real time

  - Migrate applications to app volumes

  - Migrate user profile and app settings to UEM

- PVS, MCS, Composer optional

  - Useful for storage optimization only

  - Desktops deleted and redeployed on logout

  - Deploy updates by replacing image

  - No recompose

Major Assessment Topics

Business requirements

- XenApp/XenDesktop infrastructure

- Microsoft RDSH/Terminal Services

- Published applications and desktop sessions

- Virtual desktops and applications (XenDesktop)

- Users and user requirements

- Client devices

- Supporting systems

Best Practices and Things to Consider

- **Choosing the best migration approach**
  - Is the simple way always the better way?
  - Gradual vs. one-time migrations
- **Garbage out, garbage in**
  - Moving problems from Citrix into Horizon
- **Scalability and testing**
  - Conduct pilot (live)
  - Consider simulated scalability testing

Best Practices and Things to Consider: XenApp Migration

- **Use the Identity Manager Web portal to facilitate the migration**
  1. Provides a "front end" to make XenApp migrations seamless to users
  2. First integrates Identity Manager with XenApp and then Migrate to Horizon
- **Common to have multiple XenApp environments**
  1. Typically aggregated with StoreFront, Web Interface, and/or NetScaler
  2. Identity Manager supports multiple farms

- **Should XenApp farms match Horizon RDS farms?**

  1. Application silos and load-managed server groups

  2. Putting "all the apps" on every server (is this the right thing to do?)

  3. Consolidate multiple XenApp versions on fewer Horizon RDS farms

Citrix NetScaler

- NetScaler can be configured to load balance Horizon

- Better solution: Migrate to F5 or other load balancing solution

Don't forget Horizon maximums

- Large environments using desktops and applications may exceed pod limits

Optimize the RDSH servers

- Makes a difference to implement RDS optimization best practices

- VMware Windows OS Optimization Tool

VMware App Volumes for XenApp and Horizon Hosted Applications

- Install App Volumes agent on each RDSH server/ server image

- Assign App Stacks to computers (not users)

- No support for writable volumes or user-installed apps

- Use VHD in-guest mounting for non-vSphere-based RDSH servers

- Alternate hypervisors or physical servers

- Use App Volumes to deploy ThinApps if application isolation is needed

VMware App Volumes for XenDesktop and View in Horizon

- Install App Volumes agent in each virtual desktop and desktop image/template

- Assign AppStacks to users (not computers)

- Writable volume support for user-installed apps

- Use VHD in-guest mounting for non-vSphere-based hypervisors

- Use App Volumes to deploy ThinApps if application isolation is needed

VMware Identity Manager Supports XenApp and XenDesktop

- Integration Broker syncs applications and user entitlements

- Supports single sign-on (SSO)

- Integration Broker supports Windows Server

- Integration Broker requires IIS

- SSL certificates needed on all components (Identity Manager, IB, and Citrix Farm)

- XenApp servers and XenDesktop Delivery Controller require Citrix PowerShell Remoting

- Deploy separate Integration Brokers for sync and SSO for large farms

- Identity Manager does not proxy or tunnel Citrix connectivity

Best Practices and Things to Consider: vSphere for Remote Desktop Services

VMware vSphere for RDSH Servers

- No memory or CPU oversubscription (consider memory reservations)

- 1 vCPU to each physical core and enable hyper-threading (Intel)

- Disable BIOS-level CPU power saving

- vCPUs: Less is more; four vCPUs is typically the "sweet spot"

- Do *not* exceed cores per physical CPU per RDSH server

- Scale out with more servers

- Disable DRS for RDSH servers and Horizon Connection/Security Servers

- Review and analyze memory demands for each application

- Dedicate ESXi hosts for RDSH

- Host connection servers and other support infrastructure in management block and virtual desktops in separate desktop block

Storage for Virtualized RDSH Servers

Not as IO-intensive as virtual desktops

- However, still plan for login storms and optimal performance

- Conduct assessment to determine IOPS footprint and validate with pilot

Network infrastructure

- Multiple application and desktop sessions from client requires more bandwidth than a single connection to a virtual desktop

- Be sure to assess network performance, especially the WAN

- Tune and optimize Blast Extreme and PCoIP

- Optimize the virtual desktops and RDSH servers

Tools to Help Simplify the Migration

- Automatically creates Horizon RDSH farms and application pools

- Can also migrate to existing RDSH farms

- Option to verify applications on each RDSH

- Supports Citrix XenApp 5.0, 6.0, and 6.5

- Also supports mixed 4.5/5.0 farms

- Requires Oracle Java JRE v8

- Install tool on one host in each XenApp farm

VMware Operating System Optimization Tool

- Automates optimizing virtual desktops and Microsoft RDSH servers for VDI

- Based on Microsoft and VMware operating system best practices

- Tunes Windows and disables unnecessary services and features

- Templates for Windows and Windows Server

- Templates can be customized

- Supports local and remote systems

Citrix Migration and Enhancement Driving Factors:

- XA and XD end of support

- Citrix and XenApp "fatigue"

Solutions for Both XenApp and XenDesktop

- Integrate/enhance XenApp

- Migrate/replace XenApp

- Migrate/enhance XenDesktop

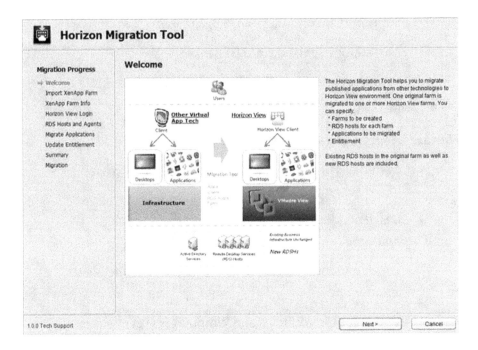

Professional Services

- Migration assessments

- Plan and design services

- Infrastructure upgrades

Tools to Help

- Assessment tools

- Migration tool

- Optimization tool

Integrating VMware tools into existing IT environments requires careful planning and execution to maximize their potential while minimizing disruption. VMware offers a suite of tools such as vSphere, NSX, and vSAN, which are commonly used for virtualization, networking, and storage. The following is a detailed breakdown.

Steps for Integration

1. **Assessment and Planning**

   - **Inventory current infrastructure**: Document hardware, software, network configurations, and workflows.

   - **Define objectives**: Establish clear goals for virtualization, such as reducing costs, improving scalability, or enhancing disaster recovery.

   - **Compatibility check**: Verify compatibility with existing hardware and software (e.g., hypervisors, storage solutions).

2. **Environment Preparation**

   - **Hardware readiness**: Ensure that servers meet VMware's hardware requirements.

   - **Network configuration**: Design VLANs, IP schemes, and firewall rules to accommodate virtualized traffic.

   - **Backup existing data**: Protect critical data by creating full backups of systems that will be virtualized.

3. **VMware Tools Deployment**

   - **Install Hypervisor**: Deploy VMware ESXi on designated hosts.

   - **Set up management tools**: Install and configure vCenter Server for centralized management.

   - **Integrate with storage and networking**: Connect to storage (e.g., SAN/NAS) and configure virtual networks using NSX or similar tools.

4. **Testing and Validation**

- **Pilot deployment**: Begin with noncritical workloads to test performance and compatibility.

- **Monitor performance**: Use VMware tools like vRealize Operations to analyze resource usage.

- **Fine-tune configuration**: Adjust virtual machine (VM) settings, resource allocations, and network policies.

5. **Migration and Scaling**

- **Workload migration**: Use VMware vMotion or third-party tools to migrate workloads with minimal downtime.

- **Scaling up**: Add additional hosts, storage, or network capacity as needed.

6. **Training and Documentation**

- **Train IT staff**: Provide training on VMware tools and best practices.

- **Document configurations**: Maintain detailed records for troubleshooting and scaling.

Common Challenges and Solutions

1. **Compatibility Issues**

- **Challenge**: Legacy hardware or software may not support VMware tools.

- **Solution**: Use VMware's Hardware Compatibility List (HCL) and consider hardware upgrades or using compatibility layers.

2. **Complex Network Integration**

   - **Challenge**: Configuring virtual networking can conflict with existing network policies.

   - **Solution**: Work with network administrators to map existing policies and use NSX for seamless integration.

3. **Performance Bottlenecks**

   - **Challenge**: Virtualized environments may introduce resource contention.

   - **Solution**: Regularly monitor performance using tools like vSphere Performance Charts and optimize resource allocations.

4. **Data Migration Risks**

   - **Challenge**: Risk of data loss during migration.

   - **Solution**: Test migrations in a lab environment and ensure robust backup procedures.

5. **Staff Resistance or Skill Gaps**

   - **Challenge**: IT staff may resist change or lack expertise.

   - **Solution**: Provide training and highlight benefits like reduced manual effort and improved efficiency.

6. **Licensing Costs**

- **Challenge**: VMware licensing can be expensive.

- **Solution**: Start with a phased rollout and evaluate ROI to justify costs.

Best Practices

- **Automate repetitive tasks**: Leverage VMware automation tools like vRealize Automation.

- **Regular updates**: Keep VMware software updated for security and feature enhancements.

- **Security integration**: Use VMware Carbon Black or integrate with existing security tools for comprehensive protection.

- **Monitor continuously**: Use tools like VMware Skyline for proactive issue detection.

By addressing these considerations and challenges, IT teams can integrate VMware tools effectively and unlock the benefits of virtualization. Let me know if you'd like deeper insights into any specific area!

Summary

If you take away with one thing from this chapter, it's that VMware must become a strategic pillar of each of your mobility programs. VMware has been working hard with the rest of the Android ecosystem to provide the most secure and comprehensive management solution on any mobile platform. I encourage all of you to really dig into Android management by leveraging additional resources and also explore migration from Citrix to VMware for a single-point solution.

CHAPTER 12

Virtual Desktops and Networking

In this chapter, we will go through the different network architecture and basic troubleshooting for a VMware Horizon deployment. We will address some basic networking scenarios along with how VDI works across LAN and WAN environments. We will also address some day-to-day challenges and the steps to resolve them.

Some Common Questions

- **Will I need to upgrade my network?**

    - No upgrade required. You probably have the equipment and features you need.

- **Will VMware View work on my WAN?**

    - Absolutely. WAN and wireless access are great when properly deployed.

- **What about voice and video?**

    - Supported. VMware View and Cisco are designing voice and video support together.

© Ajit Pratap Kundan 2025
A. P. Kundan, *Intelligent Automation with End-User Computing Solutions*,
https://doi.org/10.1007/979-8-8688-1312-2_12

- **What about security?**

  - Virtualization-aware networking can help provide superior security from end to end.

- **How does VMware help me?**

  - View offers a suite of features that helps you deliver desktops, even over challenging network connections.

Platform: VMware vSphere for desktops

Management: VMware View Manager; VMware View Composer; VMware ThinApp

User Experience: PCoIP; print; multimonitor display; multimedia; USB redirection

VMware View: Network-Related Features

- View security server

- PCoIP protocol

- View UC APIs

Data Center Considerations

- Desktop and network security

- Server load balancing and connection optimization

Endpoint Considerations

- Rich media: voice and the hairpin effect

- Network access and wireless connections

- Power over Ethernet + thin/zero clients

The Network In Between: WAN and LAN

- Bandwidth requirements

- Congestion management

- WAN optimization

View Security Server Access with PCoIP Support

Connection Sequence

- The user connects to the View Connection Server and authenticates.

- When a PCoIP desktop is selected, the PCoIP protocol goes to the Security Server.

- If the PCoIP session is on behalf of an authenticated user, it is forwarded to the correct desktop.

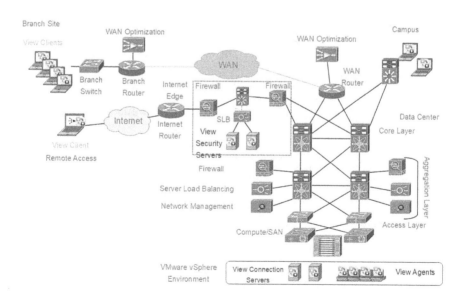

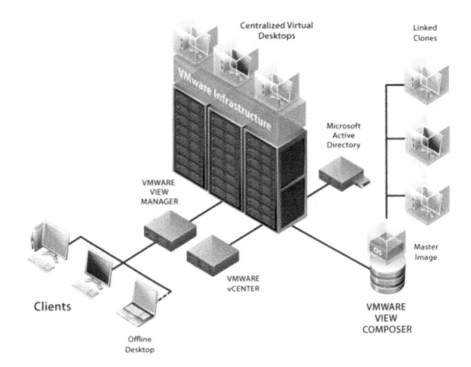

New Improved PCoIP: deploy it on a WAN near you!

WAN Performance

- Up to 75% reduction in bandwidth equals lower capex

- Increased network user density

- Configurable settings to match your LAN and WAN conditions

New Server Offload Card: Teradici

- Better performance

- Better server consolidation (equals lower capex)

Enabling partner solutions

- UC Softphone runs inside View desktops with support for VoIP and Video telephony

- New solution enables rich communications with View desktops without sacrificing scalability and user experience

- View with UC addresses previous problems of scalability, QoS, and media hair pinning

Value Proposition

- Preserve investment in existing telephony/UC infrastructure

- Seamless integration between existing infrastructure and virtual desktop model

- Provide best (voice and video) user experience

- Interactive Media Services API shared with UC partners

- API provides framework for UC partners to integrate rich media services (voice and video) within View Hosted Desktops

- Find Me Follow Me Workspace with virtual desktops and telephony enables flexible working models

Provides Virtual Access Layer Security

Distributed virtual switching with an advanced access layer features such as QoS, NetFlow, etc., to ensure a high quality of experience.

DHCP Snooping: Acts like a firewall between untrusted hosts (View Agents) and trusted DHCP servers and helps prevent the VM from acting as an unauthorized DHCP server

Dynamic ARP Inspection (DAI): Validates ARP requests and responses, uses DHCP snooping bindings, and helps prevent ARP-poisoning-based MITM attacks

IP Source Guard (IPSG): Filters traffic on vEthernet interfaces, permits traffic only where IP and MAC address match (DHCP bindings/static), and helps prevent a VM from spoofing the IP address of another VM

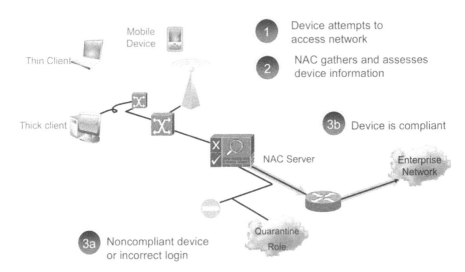

Network-Based Layer 2 Security

DHCP Snooping: Prevents malignant VM from acting as an unauthorized DHCP server

> **Dynamic ARP Inspection (DAI):** Prevents MITM
>
> **IP Source Guard (IPSG):** Prevents IP spoofing
>
> **Port Security:** Prevents Mac spoofing
>
> **Media Embedded in Display Protocol**
>
> **Hairpin effect**: Causes undesirable results
>
> **Monolithic data flows**

- Voice/video in the display protocol

- Media flow goes all the way back to data center and back

- Heavy processing on virtual desktop in data center

- Bandwidth explosion

- Display protocol and possible endpoint become unstable

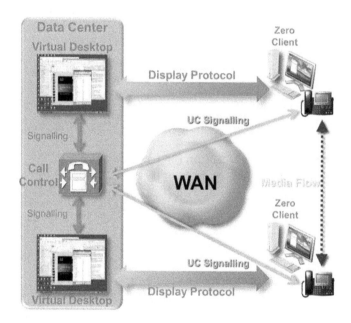

Simplify

- No QoS, CAC, and optimal codec selection for voice/video traffic inside the display protocol

- Hairpinning and high CPU utilization on HVD and the endpoint

- Wasted WAN bandwidth in cases where the two video phones are in the same branch

- Loss of Survivable Remote Site Telephony (SRST) when voice and video traffic is inside display protocol

The compression provided by display protocol is not optimal. Special codecs for voice and video have proven quality metrics that can't be achieved by display protocols.

WAN's Effects on Users Experience: Each "new" copy streamed for each additional DV client results in branch WAN bandwidth overruns.

- Capacity planning

- QoS

- Alternative paths

- WAN optimization

- Bandwidth consumption ("it depends")

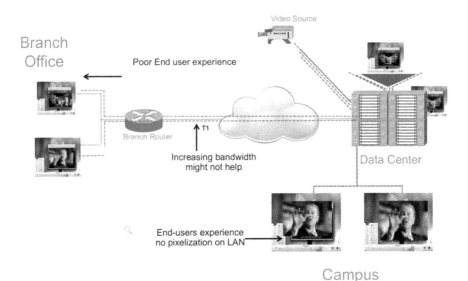

Bandwidth Consumption Is Variable

- View radically improves bandwidth profile with PCoIP

- Depends on workload, display protocol, other features (USB, etc.)

- Not every connection yields the same consumption rate

- Tunnel mode connections make View sessions look like browser traffic

Latency and Jitter Could Be Your Worst Enemies

- "Bursty" traffic and unpredictable packet arrival can have significant impact on user experience

- Dependent on protocol and tunnel mode (TCP versus UDP)

- View improves resiliency

Best Practices Help

- Capacity planning

- End-to-end QoS

- Path optimization

- WAN optimization

- Predicting/mitigating network impact of view/HVD

Capacity Planning Is Important

- Thorough testing with live applications and users (or simulated loads) is critical

- Cisco uses live data or the Reference Architecture Workload Code (RAWC) tool from VMW to form estimates

- Longer-term analysis provides superior results

WAN Optimization Is a Great Asset

- Provides compression and reduction of redundant data

- Regardless of protocol, WAN optimization improves WAN performance

- Classic examples: file share, printing at remote offices

QoS Planning Is Key

- Helps manage congestion and capacity concerns

- Must be configured from end to end

- Best practices for traffic marking included in references at end of session

- Performance routing can also improve performance on congested links

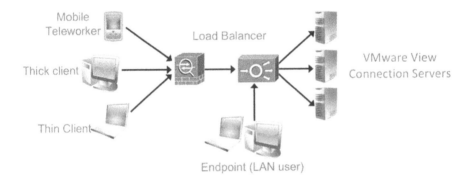

SLB SSL Deployment Options for View

1. SLB only: Performs L4-7 SLB but offers no SSL services

2. SLB + SSL End-to-End: Option1 + SSL termination (Client) and SSL initiation (Server)

3. SLB + SSL Termination: Option1 + SSL termination
    (Client) and HTTP to server

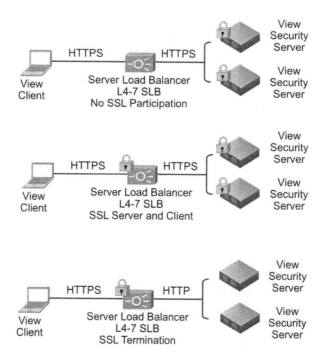

Deploying QoS for VMware View

QoS for View allows for differential treatment of View traffic.

QoS policies can protect View traffic from less important flows (e.g., voice versus CIFS).

Classification is key to place the traffic in the correct priority queue.

• Classification can occur in VMware or Cisco components in the virtualized infrastructure.

• Direct mode is ideal for proper classification.

Tunnel mode: HTTP/HTTPS flow. Classification on source/destination (i.e., View CS or Agent IP address(s)) can help distinguish flows.

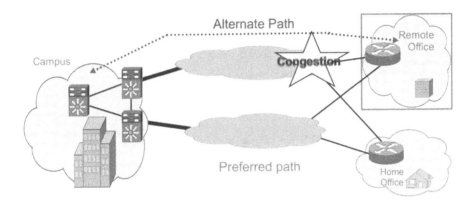

Performance Routing (PfR)

- Uses defined metrics to determine the best path
- Can route around impacted links
- Improves utilization of network for all users

Troubleshooting: Looking Under the Hood

Problem:

- "I can't even connect to View."
- "I get disconnected randomly!"
- "Why is the display so bad?"
- "Why is my desktop not available?"
- "I'm seeing an error in View. What does it mean?"
- "vCenter is reporting an error."
- "My desktop is slow."

Identifying the Problem Domain

- View Client

- Network

- View Manager

- View Composer

- vCenter Server

- Compute

- Storage

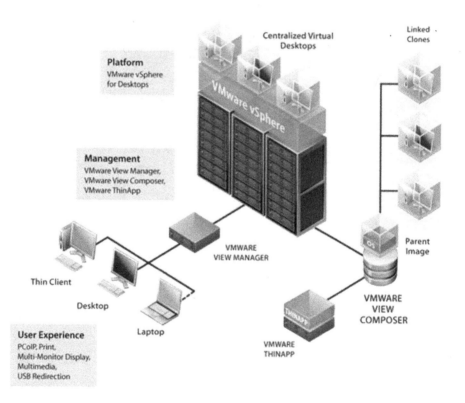

Troubleshooting Keys

- Check the View Administrator Dashboard.

  - Shows system health and any issues.

- Use the Event Database for initial troubleshooting.

- Understand the client connection process.

- Check the connection broker logs.

  - Match and filter the SessionID, user, and FSP.

- Check the view agent logs.

- Check the View Desktop PCoIP logs.

- Use kb.vmware.com.

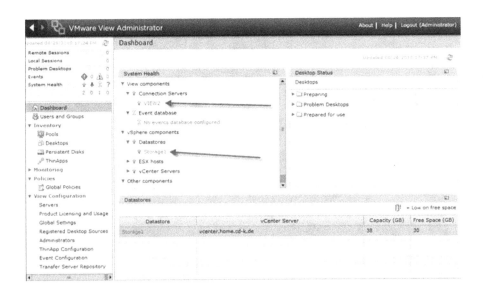

Domain 1: Failure to Communicate

- Client connectivity issues

- Common issues

- View Client can't connect

- Logon failure

- Black screen

- Poor quality display

- Randomly disconnected session

Where to look:

- Connection Broker logs (client logs)

- Event database

What to look for:

- (**Client connects**) *DEBUG <TP-Processor1> [XmlRequestProcessor] (SESSION:4C3D25DD5 E93343097E655628D96EEC7) added: configuration*

- (**Client sends user/pw**) *DEBUG <TP-Processor3> [WinAuthFilter] (SESSION:F2C008BAA8E1B2B19F35A E906F6BE79D) Attempting authentication against AD*

- (**Broker authentication**) *DEBUG <TP-Processor3> [WinAuthFilter] (SESSION:F2C008BAA8E1B2B19 F35AE906F6BE79D administrator) Attempting to authenticate user 'administrator' in domain 'MPRYOR'*

- (**User has authenticated to Broker**) *INFO <TP-Processor3> [ProperoAuthFilter] (SESSION:**4C3D25 DD5E93343097E655628D96EEC7**) User MPRYOR\ administrator has successfully authenticated to VDM*

- Event Database: BROKER_USERLOGGEDIN

User Experience Issues

- **Black screen of display**

    - PCoIP port blocked by firewall (TCP and UDP 4172) or SVGA Driver issue

    - pcoip_server/client logs - C:\Users\All Users\ VMware\VDM\logs

        - *Error attaching to SVGADevTap, error 4000: EscapeFailed*

        - *MGMT_SCHAN :scnet_client_open: tera_sock_ connect returned error 10060 - Connection timed out!*

- **Poor quality display**

    - Could be a bandwidth or latency issue

    - Pcoip_server logs report

        - *VGMAC :Stat frms: **Loss=0.45%/0.21%** (R/T)*

        - *MGMT_PCOIP_DATA :BW: Decrease (loss)* ***old = 234.9982 new = 176.8438***

- **Randomly disconnected session**

    - Often not random

    - 15 minutes after established, WSSM process hasn't started on the desktop

    - PENDING_EXPIRED View Agent logs

- Sometimes caused by daisy-chaining the GINA

Domain 2: Desktop Not Available

- **Common issues**

  - No desktop available

  - Pool provisioning issues (customization)

  - Agent not communicating with broker

  - User stuck at the desktop login screen (SSO)

- **Where to look:**

  - Connection Broker/View agent logs

  - Event database

- **What to look for:**

  - Broker returns list of desktops available to client

  - *DEBUG <TP-Processor1> [DesktopsHandler] (SES SION:**4C3D25DD5E93343097E655628D96EEC7;** F26473C8A0097F3B44A72627F9DD77CF) For user **[S-1-5-21-1850196109-1332905866-3940055309- 500]** and pool [cn=test2,ou=server groups,dc=vdi, dc=vmware,dc=int] DesktopTracker returned 3 guest DNs*

    - **What to look for:**

    - The client requests a desktop.

    - The event database is BROKER_DESKTOP_ REQUEST.

    - The broker allocates a session to the user.

- *DEBUG <TP-Processor3> [FarmImp] (SESSI
  ON:**4C3D25DD5E93343097E655628D96EE
  C7**;F26473C8A0097F3B44A72627F9DD77CF)
  newManagedSession - allocated new managed
  session on cn=f6eb929f-0a57-47b6-8b31-0734
  ab08f7f4,ou=servers,dc=vdi,dc=vmware,
  dc=int for application CN=test2,OU=
  Applications,DC=vdi,DC=vmware,DC=int
  for user CN=**S-1-5-21-1850196109-
  1332905866-3940055309-500**,CN=Foreign
  SecurityPrincipals,DC=vdi,DC=vmware,DC=int*

- The event database is BROKER_MACHINE_
  ALLOCATED.

- The Broker attempts SSO.

- *DEBUG <TP-Processor3> [FarmImp] (SESSION:
  4C3D25DD5E93343097E655628D96EEC7;
  F26473C8A0097F3B44A72627F9DD77CF)
  Using domain for SSO: MPRYOR*

- The user won't be logged on to the VM
  without this!

- **What to look for:**

  - The broker starts a session on the VM.

    - *DEBUG <TP-Processor3> [DesktopSessionImp]
      (SESSION:**4C3D
      25DD5E93343097E655628D96EEC7**;F2*

    - *6473C8A0097F3B44A72627F9DD77CF)
      startSession - sending StartSession message*

- The agent responds.

  - *"DesktopManager got a StartSession message"*

  - *"startSession added portal logon for user **MPRYOR\administrator"***

  - Event Database: AGENT_PENDING

- The client attempts to connect to the VM. The agent starts PCoIP.

  - *INFO <logloaded> [MessageFrameWork] Plugin 'wswc_PCOIP - VMware View Client PCoIP Interaction Handler' loaded, version=x.x.0 build-xxxxxx, buildtype=release*

- The client connects to the VM (agent).

  - *"PCoIPCnx::OnConnectionComplete Begin (PCOIP)"*

  - *"WTS_SESSION_LOGON"*

Event Database: AGENT_CONNECTED

- **What to look for with pool provisioning:**

  - Desktops not available due to provisioning error

    - Check the View Administrator for the pool status

    - Check the event database.

      - BROKER_PROVISIONING_ERROR_*

- Desktop not available due to customization

  - Check Desktop status: AGENT UNAVAILABLE

  - Check View Dashboard

    - Desktop Status > Preparing Desktops OR Problem Desktops

  - Check Desktop connectivity to DNS/AD/ Connection Server

- View Composer issues associated with incorrect domain credentials

  - *FATAL CSvmGaService - [svmGaService.cpp, 116] Domain join failed Error 5 (0x5): Access is denied.*

Domain 3: Broken Broker

Common Issues

- Cannot connect to vCenter

- View Composer errors

- JMS connectivity

- ADAM replication failure

Add QuickPrep Domain

Domain Information

⚙ The View Composer database has been re-initialized or is corrupt. Please restore from your last backup or see the View Composer Administration guide for help on re-enabling View Composer for View Manager.

Full domain name:

(E.g., domain.com)

User name:

(E.g., domain.com\username)

Password:

[Add...] [Cancel]

Where to look:

- View Administrator

- Event database

- Windows event logs

- View Composer logs

- Connection Server logs

- **What to look for:**

  - ADAM replication

    - Check the Connection Broker window event logs

      - ADAM (VMwareVDMDS)-log

      - Error: ADAM Replication

- Check the ADAM replication status on the Connection Server

    - *C:\WINDOWS\adam\repadmin.exe / showrepl localhost:389 DC=vdi, DC=vmware,DC=int*

- vCenter Server Connectivity

    - Admin UI will show a RED status

    - Check the event database

        - VC_DOWN events

    - Impact provisioning and power operations

    - Check the connectivity from the connection server to vCenter Server

    - Check the credentials used to connect to the vCenter Server

        - Attempt to log in directly to vCenter using vSphere Client

What to look for:

- View Composer

    - VMs have been manually deleted; then pool/ desktop deleted

        - Causes Composer DB and VC DB to get out of sync

        - Composer thinks VM already exists

        - Orphaned VMs

        - *Desktop Composer Fault: 'Virtual Machine with Input Specification already exists'*

- JMS Connectivity

  - Split site architecture/firewall causes "split brain"

  - View Dashboard shows RED status

  - Connection server logs

    - Tracker REJOIN messages: JMS connectivity

    - Tracker RESYNC messages: messages being delayed

Domain 4: Why is my desktop so slow?
Common Issues

- Storage IO bottleneck

- Memory contention

- CPU contention

- Network issues

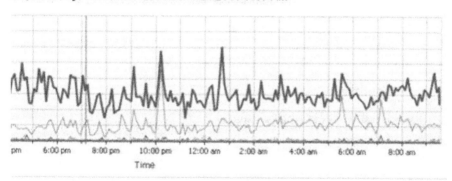

isk, Past day: 6/23/2010 9:35 AM - 6/24/2010 9:35 AM

Counter	Units	Latest	Minimum	Average	Maximum
Disk Read Latency	Millsecond	244.00	33.00	304.68	606.00
Disk Write Latency	Millsecond	96.00	26.00	108.57	462.00
Disk Read Latency	Millsecond	8.00	2.00	9.13	51.00

Where to look:

- vCenter Server

- 3rd Party Tools?

What to look for:

- CPU

  - Cluster/host utilization

  - VM utilization

  - VM %READY Time (ESXTOP)

- Memory

  - Host utilization

  - VM utilization

  - Swapping/ballooning

- Storage

  - VMs per VMFS LUN (SCSI reservations)

  - Disk Read Latency < 20ms

Action points to resolve these issues:

- Read the product documentation.

- Double-check your configuration!

- Check kb.vmware.com for your issue.

- See http://communities.vmware.com.

- Run Support.bat to extract the logs.

  - Notice the diagnostic tests that run.

- Submit a support request.

Summary

You have to first understand where the issue may lie. Is it the client? The agent? The server? The composer? VC? ESXi? Know the problem domains. You also have to check the View Dashboard and event database regularly to identify the issue and must know what a successful connections looks like. Then, we can go through the logs and get the required help from the help contents.

Consider the entire system from end to end:
Consider QoS, security, and other considerations.

You probably already have most of what you need:
Following best practices will help you get the most out of your existing infrastructure.

View works well on the WAN or via wireless:
Understand network usage for the best results.

Voice and video support is no longer optional:
Network design must supply rich media features.

Index

© Ajit Pratap Kundan 2025
A. P. Kundan, *Intelligent Automation with End-User Computing Solutions*,
https://doi.org/10.1007/979-8-8688-1312-2

GPSR Compliance
The European Union's (EU) General Product Safety Regulation (GPSR) is a set
of rules that requires consumer products to be safe and our obligations to
ensure this.

If you have any concerns about our products, you can contact us on

ProductSafety@springernature.com

In case Publisher is established outside the EU, the EU authorized
representative is:

Springer Nature Customer Service Center GmbH
Europaplatz 3
69115 Heidelberg, Germany